IN HIS **SIGHTS**

IN HIS **SIGHTS**

ONE WOMAN'S STALKING NIGHTMARE

Kate Brennan

NEW YORK • LONDON • TORONTO • SYDNEY • NEW DELHI • AUCKLAND

HARPER PERENNIAL

This is a work of nonfiction. The events and experiences detailed herein are all true and have been faithfully rendered as I have remembered them, to the best of my ability. All names and identifying characteristics of certain individuals have been changed in order to protect their anonymity.

A hardcover edition of this book was published in 2008 by Harper-Collins Publishers.

P.S.™ is a trademark of HarperCollins Publishers.

HarperCollins books may be purchased for educational, business, or sales promotional use. For information please write: Special Markets Department, HarperCollins Publishers, 10 East 53rd Street, New York, NY 10022.

FIRST HARPER PERENNIAL EDITION PUBLISHED 2009.

Designed by Leah Carlson-Stanisic

Library of Congress Cataloging-in-Publication Data is available upon request.

ISBN 978-0-06-145162-1

09 10 11 12 13 DT/RRD 10 9 8 7 6 5 4 3 2 1

A handful of people have lived through this story with me. While many people in my life either ignored what was happening to me or passed off the stalking as an exaggerated drama—a love that wouldn't let go or something of my own making—a number of relatives and friends recognized it as a man's rage turned to madness. They stood by me and did whatever they could to help me stay safe and sane. And when I decided that telling my story might help other women figure out how to stay safe, they encouraged me to weigh the risks, and then to write.

Acknowledging by name those still living could put them in danger, because some stalkers, like mine, are too sick and too arrogant to be afraid. So the following will have to do:

You know who you are.

To each of you, hand over heart, thank you.

I'll never forget.

I'LL BE WATCHING YOU.

The Police

Prologue

You seldom choose the circumstances that offer meaning to your life. Given a list of options, stalking isn't one I'd ever pick. But once that was my reality, I saw two basic choices—walk straight through or shy away. My nature is to walk straight through the hard things—grief, sorrow, fear, doubt, anger, whatever presents itself. I've always believed in the power of the other side of pain, so I don't allow myself to run from it.

Facing your demons, taking responsibility for your choices, learning from your mistakes: that's the kind of person I respect and aspire to be. So being stalked by an ex-lover requires me to examine how I managed to love such a man. My stalker may have picked me, but I picked him, too. I picked him, I lived with him, and when I left him, I intended to remember him in only the vaguest way, the way you recall a movie that didn't live up to the hype.

I thought it would be a simple matter of walking away and taking stock: tuck back into myself, consider why I chose him, face my frailties and failings, and then, and only then, step into the future—wiser and more whole than the day I met him. That had always worked for me in the past.

Turns out, this time would be different. You can do all the psychic and physical separation you want, but there's no getting away from someone who wants to remind you he can mess with your life anytime he wants. Paul isn't a man who tolerates being left. His desire to control me didn't vanish just because I tried to. In fact, the stalking has lasted far longer than our life together. So every choice I make, every moment of turning, is filtered through one simple fact: my stalker is still alive. Which is why you won't know my real name. But you will know my psyche, for I intend to offer it bare as a licked bone.

Being stalked thrusts you into the muck of someone else's life, which is how I felt when I was with him, so it was a surprise that I wasn't, after all, free of him when I walked out his door. How could I know that leaving this man would give new meaning to the concept of afterlife? No matter what else happens, my life will always be divided into three parts: before him, with him, after him. Not my preferred life markers. But there it is. It's what I got when I walked away.

I had forty-one years before him, nearly three years with him, and it's been more than thirteen years since I left him. It took me more than two years to see that leaving him was not the same as getting away from him, and that his harassment was, in fact, stalking.

Over the years, therapists have assured me that they, professionals who are paid to figure people out, were fooled, just as I was, by this man's charms. I'm only somewhat consoled by such assurances, because they don't erase the inevitable questions: How could I have loved someone so capable of residual hate? How did I allow myself to get sucked into his perversion? How did I manage to get away? How do I stay safe now? And most important of all: How do I keep sane, not ever knowing if the stalking is over?

The answers are complicated, but the truth is simple: it all flows from the currents of my past.

It took living with the man who became my stalker to realize that life with my family had left me with such a high tolerance for cruelty I couldn't recognize perversion when I saw it. And when I did start to see it, I was so accustomed to thinking that sick people get well and that I could survive anything, I didn't know when to quit hoping. I didn't know when to quit being strong and patient and kind.

Some women are raised to believe men can become their best selves if they're not left to their own limitations. We're bred to believe in the power of redemption. It took time—too much time—for me to realize that picking someone who needs you, who's less whole than you are, is the easiest way to keep from seeing yourself clearly. It offers ready distraction from your own damage.

I thought that avoiding active alcoholics and working on my own frailties would be my salvation. I also thought if I understood enough about the man I loved and was a steady force of love for him, it would all come right in the end.

Turns out, I couldn't have been more wrong.

I startle awake. A gunshot? I hold my breath and strain to identify the noise in the ensuing silence. Nothing. Was it a car backfiring on the highway? A hunter downing a deer on the island? Maybe one or the other. Maybe something more ominous. I'm never sure which way to turn, toward the ordinary or the terrifying.

I lie still and focus on my surroundings, work my legs toward the edge of the bed and ease into position. I've taken to wearing sweats and a long-sleeve T-shirt to bed instead of pajamas in case

I have to leave suddenly. Running or taken, either way, it makes me feel a little less vulnerable. I edge my arm from beneath the cool sheet, slide the cell phone from under the spare pillow. I practice the drill just in case: my fingers slide over the keys: 9-1-1. 9-1-1. I remind myself to breathe—slowly, silently. I remain as still as possible.

I lay my head back against the pillow to keep my neck from locking and play back the moments before going to bed. I see myself checking all the doors, the windows, the security system. Once. At least twice. Perhaps, like many nights, more than that. I visualize the routes I'll take if someone comes through the front door, through the back door, through a window. I wait for another sound that will tell me which way to run.

I picture a face I've never seen before, the stranger he will have sent. I imagine the place I'll be taken. Dirty, dark, remote, somewhere as foreign to me as my assailant. I remember to breathe. I strain for the next sound, hoping it will not come. But also wondering if it would be a relief from this endless vigilance.

I wait for the pink of morning to bleed into the night sky. Although I don't believe the dawn cares for humans one way or the other, only at first light will I relax enough to fall back to sleep.

I feel safest in the light, but I am drawn to the cloak of invisibility darkness offers.

Going to bed, for a walk, to a movie—such ordinary things. Unless you're being stalked. Then everything feels risky.

As I do in every public place, when I walk into a movie theater, I enter slowly, scanning the seats. Is he here? Is someone he knows here? Before I settle into the room, any room, I check my gut. Do I feel safe?

My favorite spot in movie theaters used to be two-thirds back, on the right. Not too close to the screen, not too far away. Now I sit in the back row, so I can keep an eye on the room. And before I give myself over to the story on the screen, I make sure I know where the exit door is. If there's only one door, I usually leave, and wait for the DVD.

When I do stay, I move in and out of the film to check for late arrivals, and minutes before the end, as the movie closes in on itself, I make my final descent from the screen story to my own. I plan my exit. How to be seen by the least number of people. How to get out the fastest. That's the goal.

I pull my hat down over my eyes and focus on the exit door tucked behind the screen. In warm months, I feel more exposed. Midwesterners aren't crazy about eccentrics. They're most comfortable with conformity. And a woman wearing a hat in a movie theater when it's above zero doesn't fit their definition of normal. People tend to stare. So, like a child who believes she's invisible when she covers her own eyes, I put on my large sunglasses and tell myself no one will recognize me.

As the credits begin to roll, I head for the front corner of the room. In seconds, everyone else will be moving in the opposite direction. The casual eye could assume I have something to hide. And actually, I do: myself. If I hadn't learned how to hide, I wouldn't have lasted this long.

I'm good at figuring things out. I'm good at figuring people out. And I'm good at trusting my gut. Years of being stalked have made me an expert at following my instincts, which isn't easy, given that a stalker's goal is to suck you into his vortex; when your world spins out of control, it's easy to lose your balance. A constant state of readiness is the only way I know to right myself, again and again and again.

But it wears me out, and the possibility of despair is never more than an arm's length away. That's on good days. On bad days, despair sits on my shoulder waiting for the slightest sign of weakness so it can wrap me in its embrace.

Memories are closer still. They crawl around inside me, hookworms pleased at so much strength to sap.

PART I

1991–1994

Chapter 1

It's my last full day in Haworth. I get up early and rush through breakfast at the B and B, passing on a cooked meal, settling for coffee and toast with homemade marmalade. I want to be at the library as soon as the staff opens the doors. Running through the mental list of manuscripts I want to review this morning, I lean against the stone wall on the edge of the parking lot. Just beyond the Georgian parsonage, white cumulus clouds roll over the moors, casting shadows across the grass and heather. Before this trip I didn't know that heather was an evergreen, its whorled leaves and waving petals so different from the spruce and pine and fir at home. From a distance the moors are a soft world of purples and browns and greens, but up close they reveal rough terrain and reservoirs full of water so cold it can kill.

"You're here early," says the head librarian as we pass through the shop on the way to the library. She looks the part. Somber dark suit. Serious glasses. Even her long mahogany hair is wrapped tight in a bun. But over the past two weeks, I've had a window into her evenings and weekends. Not so serious then. I like that—second impressions that pleasantly surprise.

It's already hot and windy outside, but inside the library it's dry and still and cool, to accommodate the manuscripts. Here dead people count more than the living. So each day I throw a sweater and pair of socks in my backpack on top of the notebooks and pencils. No pens allowed. I put on the cotton sweater now and reach for the white gloves I'd left on the long wooden table yesterday afternoon. The library, available to scholars by application, is housed in the part of the Brontë Parsonage that was once the kitchen where Emily reigned—when she wasn't haunting the moors, probably trying to escape Charlotte's hovering.

But it isn't enigmatic Emily or stern Charlotte who brought me to this library. Rather it's Anne, the least known, least favored, least published Brontë whose secrets I'm hoping to unlock. Charlotte painted her as shy and weak, but that doesn't do justice to the woman who wrote *The Tenant of Wildfell Hall*, a telling story of alcohol and drug abuse, of rape and escape.

At the end of each afternoon these past two weeks, I've left a note detailing the materials I'd like to review the next day. A selection from last night's list is neatly stacked on the end of the table. I pull on the gloves and pick up the first letter. I'm the only one in the room; the head librarian has gone to retrieve another manuscript I've requested, and the rest of the staff haven't arrived yet.

I'm surrounded by books. The oldest and rarest are locked away for safekeeping, available only by written request, but at this stage of my research, I've had to adopt a strict system. I'm tempted to stay another week, but I've promised a friend I'll be home in time for her fortieth anniversary party. So instead of diving headlong into each and every book, I make lists—of books and manuscripts, of copies I want made. I force myself to save the reading for home, or for my next trip here. But I do take a few moments to study the minute script of one letter. In April 1849, the month before she died, Anne wrote Ellen Nussey, a family friend:

> I have no horror of death: if I thought it inevitable I think I could quietly resign myself to the prospect. . . . But I wish it would please God to spare me not only for Papa's and Charlotte's sakes, but because I long to do some good in the world before I leave it. I have many schemes in my head for future practise—humble and limited indeed—but still I should not like them all to come to nothing, and myself to have lived to so little purpose. . . .

This doesn't sound like the "resigned" Anne that Charlotte would have the world believe in, a young woman "thankful for release from a suffering life." Charlotte keeps to one line impossible for Anne to argue from the grave: a girl who "from her childhood seemed preparing for an early death."

I don't buy it. Ever since I came across a reference to the letter Charlotte wrote Anne's publisher in response to his request to posthumously reprint *The Tenant of Wildfell Hall*, I've suspected Charlotte manipulated her youngest sister's image. I want to dig around to see if my hunch is correct. But Charlotte destroyed so many of her sister's papers, the space between her portrayal and Anne's own words has proven as wide and hard to navigate as the moors outside these doors.

Wildfell Hall "hardly appears to me desirable to preserve," writes Charlotte to Anne's publisher in 1850. She refers to a book that sold so well a second printing occurred six weeks after the first, and the publisher was requesting permission for a third.

The idea of contributing to the scholarship that could restore Anne's literary reputation is thrilling. As a journalist, I'm known for my interviewing skills, but I prefer sorting through the secrets of the dead to tracking down those of the living. I'm most comfortable alone, researching and studying people who talk to me in their writing. Listening in person wears me out, I suppose because I have little patience for posturing, and evasion.

Some days in Haworth I don't even break for lunch. Others I come out of myself enough to join one of the librarians for a walk through the narrow town built around steep, uneven cobbled streets, and a quick lunch or tea. Sometimes I do it just because I know it's good for me. But not on my last day, when I step outside at noon and settle in with an apple on a wood-slatted bench in front of the parsonage. I imagine the Brontë sisters, dark wool hems swishing against the wide stone steps, coming and going, to church, the village, the moors. I've spent so much time with these women in my head, I can almost see them. Back inside, I work through the afternoon, until the staff is ready to leave, then pack up all the papers I've gathered and say my good-byes. On my way through the shop, I buy several copies of *Poems by the Brontë Sisters*, gifts for a fund-raising event I'm helping plan back home: Claire Bloom giving a dramatic reading of *Jane Eyre*.

The benign morning sky has been invaded by a dark bank of threatening clouds. But to the east, it's completely sunny. I stop in my room before I set out for my last walk on the moors. Polly, who owns the B and B, and I plan to go as far as the reservoir today, so I pick up my day pack, already stuffed with a rain jacket, bottle of water, and ordinance map. I'm prepared for a change of weather. The books and movies don't lie about the moors. It's easy to get lost, to disappear.

In London the next day, I take the tube to Charing Cross and walk up St. Martin's Place to the National Portrait Gallery. My goal is to see the only portrait that exists of the three Brontë sisters together. It hangs in a second-story room along with portraits of other British writers, the Brownings, George Eliot, Tennyson.

Although the canvas is about two and a half by three feet, it's smaller than I'd imagined. I think I expected them to be life-size. It's often that way: we expect the physical reality of a found object to be equal to the emotional impact, and yet it seldom is.

The sisters all wear white collars over dark dresses. On one side are Anne and Emily, to the right is Charlotte. All three wear grave expressions, though they were only in their teens when they sat for the portrait. Charlotte's image is the most clearly defined; she seems to be in a spotlight, consigning Emily and Anne to the shadows. It's as if Branwell had studied his older sister more carefully. Or perhaps had simply known her better.

The plane makes a wide loop over the city as it prepares to land. At this point, the Mississippi flows west to east, like a snake whose lazy midsection rested before dropping south. I look down to see if I can spot my apartment building. I live on the edge of one of the city's many lakes, all of which are easy landmarks from the sky. Mine, the one shaped like a bald head with a bump on the crown, is tucked between a bean-shaped lake to the south and one shaped like a kid on a skateboard to the north.

I arrive home midafternoon, with plenty of time to shower and change for Jen and Doug's anniversary party, which is being held at the home of their oldest daughter. Since they moved in across the hall several years ago, they've become like family, especially Jen. She's the aunt every woman wants to have: intensely interested in my life without being overbearing. Her extraverted life, full of news about her family and harmless gossip about neighbors, is the perfect antidote when deadlines keep me at my desk.

It's as hot and muggy as London was. I dress in the coolest and most comfortable clothes I find in my closet: white silk blouse, white linen jacket, blue silk skirt, and sandals. No skimpy dresses for me; I have such sensitive skin I burn even in early evening. Even more pale than usual because I'm tired, I swipe SoHo on my lips: the cool raspberry color intensifies the steel-blue of my eyes. I can hear my mother's voice: When you're tired or sad, put on something bright—but not too bright or it'll wash you out. The color of her blouse or freshly applied lipstick was always a clue to whether she was having a hard time of it.

I reach the party just as evening—and, thank God, the humidity—is beginning to fall. The lawn is freshly cut and watered, damp but not wet. Dotting the backyard are green-and-white-canopied tables, covered with white linen cloths, in turn anchored by vases full of fresh flowers and candles waiting to be lit. This is a well-manicured suburban family.

Typical of most midwestern parties, nearly everyone here is paired. I'm comfortable being single except at gatherings such as this, where I feel my difference. I'm sitting with good friends of Jen and Doug's, two couples who recently moved into our building. One of the men is asking if I'd be interested in editing a family history he's just finished writing. We've agreed to meet the next day over coffee so I can see what he's got.

Not long after I arrive, Jen comes to the table and leads me to where her brother Roger stands. He's perfectly pressed, summer slacks, white shirt, navy blazer, all worn like a uniform. He's flown into town especially for the party; years earlier he'd moved out to the Southwest. Jen has two brothers and no sisters. All I know about the other brother is that his plane went down somewhere in France during World War II and his body was never recovered. I've heard a bit about Roger: the cruises, guest houses, five-star hotels, and Zagat-rated restaurants, always his treat. But it sounds as if his hospitality comes with hooks. I'm not disposed to like him.

I suppose if you have only one brother left, you hang on,

regardless. I know how awful this sounds, but having five myself, I tend to feel they're all expendable. As far as Roger is concerned, I'm always ready to be wrong, so I plan to greet him with an open mind. "I want you to meet my friend," Jen says, offering my name as if it were a gift. "I've told you so much about her."

The minute she begins the introduction, he shifts his focus beyond me. I know his type, a barrel chest full of bravado, always looking for someone more important than the person in front of him. I plan to do the polite thing, extend my hand, put on a cordial face, and tell him I'm happy to meet him.

But as my well-intended hand drifts toward him, I look into his pale blue eyes, as flat and cold as I've ever seen. I swiftly withdraw my hand and slip it into my jacket pocket. Startled by my own reaction, I manage to mouth a truncated version of the words I'd prepared. "Nice to meet you. I know how fond Jen is of you."

"Yes," he says, still looking past me. He appears practiced at letting others know how insignificant he considers them. And yet, I'm surprised by his lack of pretense in front of his sister.

After dinner, and many toasts, I mill around, searching out the handful of people I know. This brings me to a table where Doug and Jen's second son is sitting. Jack greets me warmly, then turns to the man sitting next to him and introduces me to his cousin, Paul, Roger's only child.

In a sea of other men in summer suits and light blazers, Paul stands out. He's wearing white linen trousers and a short-sleeved, bright-blue silk shirt; we could be mistaken for a matched pair. He's good-looking in a Hemingway hero sort of way—sturdy, worldly, athletic-looking. He has his father's square face, but Paul's is softened by wavy reddish hair—the kind of red that fades to strawberry blond on its way to gray. His pale blue eyes are accentuated by the color of his shirt.

Sitting next to him is a woman I judge to be a few years younger than me. She's got short, reddish-brown hair and is wearing a simple sundress that sets off her tan. She looks fit. Paul is bringing a friend, not a date, Jen had told me earlier. I wondered, but didn't ask, the point of telling me this. Perhaps it was simply conversation. She clearly adores her only nephew. Everything Jen has told me about Paul runs along two lines: his exciting life of travel and his difficult relationship with his father. She describes a man who can't break loose from a father whose psyche depends on belittling his child. It's the only flaw Jen ascribes to her brother, and the one detail I remember.

As I slide into a seat next to Jack, he asks about my trip to England. I'm delighted to be talking about my research; I don't expect people to be interested in such an obscure literary puzzle. Paul is also attentive, and his questions reveal he's as passionate about travel as I am. He's a freelance photographer; I know from Jen that he's got family money and doesn't have to work, so I'm impressed. Jack names a glossy international magazine and prods Paul to talk about his photo shoots in Africa, Asia, and South America. Jack leads Paul to his favorite: a winding road in the Colombian jungle. An ambush, guns pointed at their approach. Someone in the jeep has a gun, too, but Paul's quick thinking and fearless driving ensures they don't have to use it.

Already I can tell he has a knack for putting himself in the center of the story. Part of my brain registers desire. I'm primed for this particular model of man. I grew up on adventure stories—real ones, as well as fictional.

The woman with Paul grows quieter the longer we talk. It occurs to me, in spite of Jen's comment, that his "friend" considers herself his date. From time to time, mentioning one family party after another, he asks why we haven't met before. My response doesn't vary: "I don't know. I was there. So were you." It's true that I never noticed him, but I also say it to balance out the charm he knows he exudes. It's subtle, but it's there, in the insouciant way he sits, the way he holds his head. He bears himself like a man used to being remembered.

Jen reappears at the table and whisks Paul away to say hello to family friends. I talk to Jack for a while longer, then start my round of good-byes. After thanking Jen, I turn to Paul. "I'm happy we met," I say.

He aims his pale blue eyes directly into mine. "I'll see you again," he says. As he extends his hand for our first good-bye, I feel a jolt of certainty: my life will never be the same. Off balance, I reply, "I imagine you will." I hadn't meant to say this aloud, but I've already suspected he won't be able to resist pursuing someone who hasn't noticed him before and tells him so.

We don't need to trade phone numbers. He knows where I live.

Chapter 2

One afternoon the next week, Doug calls to tell me Paul has dropped by their apartment and wonders if he can come over.

I'm organizing the documents I brought back from England. I'll be presenting a paper at the Midwest Modern Language Association conference in a few months, and I want it to be as loaded with facts as possible.

"Now?" I ask.

"Yes," Doug says.

"Sure," I say. One of the advantages of working at home is the ease of folding personal time into work. I don't have plans this evening, so I know I can finish the project later. Doug tells Paul it's okay. In the background, I hear Jen saying good-bye, then the door closing. Just before Doug hangs up, he laughs. "He never comes to visit us," he tells me. "He stopped here only to see you."

By the time I've put down the phone, Paul is at my door. He looks relaxed, freshly tanned, as if he's just walked off a sailboat, or a golf course. He's wearing khakis and a bright red polo shirt. His lightweight hiking boots seem out of place on this hot afternoon. He looks ready for an impromptu hike, but in the city?

I pour glasses of iced tea, and we sit in the living room on the white couch that faces a wall of windows. The first time I walked into this sixteenth-floor apartment thirteen years ago, I knew I wanted to live here. It was smaller than I had hoped for, but the southern window of the living room looked out upon a four-hundred-acre city lake, and stepping onto the cerulean-blue carpet made me forget the room was bound by concrete and glass. On one wall is a birthday gift from my parents: an oil painting of waves rising like clouds from a rocky lakeshore. I've positioned the painting so the trumpeter swans are racing toward the lake outside my window. In the corner of the room is a round walnut table, piled high with books and folders. Paul asks how my research is going. I give him a brief preview of my talk, then steer the conversation his way. I don't have to ask many questions to get him to pick up where we left off at the anniversary party.

Despite the drama of his stories, Paul's manner is gentle. He is reserved yet engaged, an interesting combination. And he's witty. My second impression of this man corresponds to my first: he seems a welcome departure from the usual model of midwest manhood, all business and sports and God. Nothing against any of those in general, but as a package I find them limiting, boring even. And there's something else I see. Try as he might to hide behind a relaxed and confident manner, his charm doesn't mask a vulnerability.

As much as I'm drawn to him, I have a bad romantic track record, so I'm wary. The men I've dated aren't bad people, just garden-variety insecure or selfish men who don't want to relinquish first place in their personal lives; they aren't looking for a partner so much as a second in command. And as the eldest daughter in a large family, I've had my fill of mothering, so the idea of a man as a means of securing a family of my own has never particularly appealed to me. Hard as I try, I can't fix on a picture of myself in a lasting pair.

Even as a child I was most content when left alone, to read a book, to daydream, or to listen to the birds, trying to identify them by their songs. It was the first time I realized you could like the way a thing looks but wish it would keep its mouth shut. Take blue jays: if only they sounded as beautiful as they looked. Sitting alone in the small woods behind our house, with a book in my lap, I realized I liked it best when animals, and people, didn't reveal unpleasant surprises when they opened their mouths.

Once a relationship hits the inevitable point at which I realize I can salvage it only by diminishing myself, I prepare to leave. It's simple, really. I'm eager to encourage another's dreams. I just won't sacrifice my own. So, at the age of forty-one, I've decided that instead of taking another lover I'll enjoy deep friendships and develop an even richer interior life.

But then I meet Paul, and I can't think of a solid reason to turn him away.

Before he leaves my apartment for the first time, he tells me he's going on a vacation, hiking in the mountains, he says, though when I ask where, he's not specific. He asks if I'd like to go out when he gets back in early September. At the door, he turns and almost shyly leans my way. "I'd like to hug you good-bye," he says. "May I?" I've never met a man who asks for permission like this. I find it formal, which he isn't, and odd. I haven't decided whether I like it or not, or what it says about him. The hug lingers just long enough to be remembered but not so long that it leads to anything else. Yet I know this moment is the start of something.

Years ago I fell in love with a man I'd met while working at a law firm. Jim's parents died when he was in his early twenties, and he was an only child. I actually enjoyed his orphan status; I liked that loving him didn't mean having to fit into another whole family. What I can't love about Jim is his grasp of ethics when it comes to business. Too loose for my comfort.

Even though we stopped being a couple several years ago, the relationship remains open-ended. Every time Jim asks me to marry him, I tell him my answer has to be no as long as his business practices remain the same. But he never quits asking. And I've come to depend on his long-distance adoration. He now lives in California, so it's easy to keep the fantasy of our eventual coming together alive. "When we're old," he says, "we'll be together." I believe it. I'm comforted by his unwillingness to let me go completely, because his passion, or attachment at least, is not threatening. He simply calls once every several months to see if my answer has changed. In between the proposals, he has broken three engagements. I flatter myself that this has something to do with me. At least that's what he tells me.

Although I keep saying no, I hang on to the idea of us together. Still, I try to be open to more immediate love.

August ushers in a trio of deaths, and one marriage. I prefer family funerals to weddings. For one thing, I don't feel conflicted at funerals. When people die, you're sad in direct proportion to how much you loved them.

Weddings are another matter. When you come from a long line of alcoholics, if you're not genetically wired to be one yourself, your odds of marrying one are depressingly high. When someone marries, you're meant to feel joy. But when you watch siblings and cousins marry people who are clearly drunks, or, worse yet, are drunks themselves, well, your instinct is to grieve, which is inconvenient when you're expected to wear a big smile.

Fortunately, my family is gaining a certain degree of sobriety. So when the third of my five brothers gets married shortly after my return from England, I don't dread the event as much as usual. Still, I don't look forward to these times with my family. I prefer the quieter gatherings, when only a few of us get together rather than the whole bunch; we tend to act like civilized adults then. But put more than a handful of us in a banquet room, or someone's living room, or anywhere for that matter (all we need is a critical mass), and a primitive gang mentality kicks in. We don't make much room for introverts—much less loners—and standing apart is translated as contrariness, and differences of opinion often escalate into personal attacks. It's a shaky form of love, not really love at all. Just an unrelenting tie that won't release its hold.

I no sooner rest up from this dose of family togetherness than I'm on a plane. My mother's last surviving aunt has died, so most of us go to Chicago for the funeral. I'm gone for several days, and when I call home to check for messages, I'm pleased to find there's one from Paul. He says he'll call again when he gets back to town.

Chapter 3

About a week after I get home from Chicago, Jen knocks on my door. It's early, a few minutes before eight. I've just made coffee but haven't had time to drink it yet. I'm still in my robe, but so is she: a plush terry-cloth robe with a medallion on the breast pocket. Her thick, short gray hair is matted against her head. That alone tells me something's wrong; Jen is always pulled together. Her eyes are red, her tanned face puffy. She looks as if she hasn't slept.

I hook my arm through hers and pull her into the apartment. Jen is tall and solid in a singularly American country-club sort of way. This morning, there's nothing hearty about her: she looks lost.

She slumps down onto the couch. She's crying so hard, I can hardly understand what she's saying. Bit by bit she gets it all out. The police called early that morning. She expels the next sentence as if the words alone are fatal. "Roger's dead." They found her brother yesterday afternoon. He didn't show up for a golf game, and he didn't answer his phone when his friends called. They got worried and notified the police.

"My God, Jen, I'm so sorry." We're sitting next to each other.

I'm holding her hand, stroking it the way I do to calm my nieces and nephews when they cry. "Was it his heart?" I assume so. He'd had a heart transplant several years earlier, I recall.

"He's been murdered," she says, sitting up straight. And as if I might not have heard the word the first time, she repeats herself, "They say he's been murdered." She spills out the rest of the story in a rush, as if she's eager to be rid of it. The police found him at home, in his bedroom—strangled. It took them several hours to reach Paul.

"Paul says he was home, just not answering his phone," she says. "Why wouldn't he answer his phone?" she asks, as if I might know. "It took them hours to reach him." Her gaze drifts away to the view outside my window. Suddenly she shifts back to me. "Why wouldn't he answer his phone?" Again, as if I might know.

I want to comfort her. Over the years we've been neighbors, we've grown to love each other in that realm between family and friendship. I murmur words of comfort, but I know nothing will ward off the days and months of pain that have barely begun.

Within a few minutes, she reverts to the pragmatic self I know best and tells me their plans. Now that everyone in the family has been contacted, she, Doug, and Paul will fly out to Albuquerque this afternoon to meet with the police. Paul insists on seeing his father's body, Jen says, even though the police recommend against it. "It's the strangling," she adds, launching into a gruesome explanation of how it distorts a person's face. As she does so, her own face relaxes, as if the clinical details cancel out the reality. "But Paul insists," she repeats, as she gets up from the couch. "I don't understand."

At the door, I wrap my arms around her and hold her, my right hand rubbing her back, the way a mother comforts a child. I can think of nothing to say, except to assure her that I'll be here whenever she needs me. As I stand in my doorway and watch her walk down the hall, it occurs to me that her life will never be the same.

This is her first brush with a violent crime, but not mine. More than twenty years ago, one of my mother's cousins was found beaten to death in his Chicago apartment. They never found the murderer, which had, I always thought, everything to do with the fact that John, a man who always brought laughter to a room, was homosexual. The police speculated he'd met his murderer in a bar and either my cousin invited him home or his killer followed him. I was eighteen, old enough to gather from the phone calls my mother received that the police hadn't tried very hard to solve the crime.

I suspect Roger's murderer will be more vigorously pursued, even though Roger, too, I suspect, was a closeted homosexual. It always seemed obvious to me from Jen's stories that all the young men Roger put through college or helped set up in business were lovers he snared with money. But curiously, Jen never once acknowledged the possibility that these relationships were sexual. It's as if she had a blind spot.

Though I have no feelings for this man I'd met only briefly, I'm worried about my friend, for no matter the precise circumstances, no matter the motive, even the few details I know of Roger's death horrify me. I can't help but think the facts that emerge in the murder investigation will force her to accept a truer picture of her idealized brother, and add an extra layer of grief.

The next time I see Paul is at his father's memorial service. The vigorous man I remember has disappeared. He's dressed in a dark suit, expensive and well cut, but too large by at least a size; perhaps it's his father's. I observe this like a reporter. It isn't something you can ever completely turn off, but at moments like this, my compassionate self wants to reprimand the cool mental note taker.

I arrive at the service a few minutes late and slip into a pew toward the back of the large modern church with clerestory windows. No casket, not even an urn. The police haven't released the body yet. The church is crowded even though Roger moved from this city years ago. I plan to stay only long enough to greet Jen and Doug. They'll be busy talking to the people they haven't seen since Roger's death. Besides, I hardly know Paul. When I tell Jen I'll see her at home, though, she insists I go talk to her nephew. "He asked if you'd be here," she says, nodding in his direction.

I follow her eyes. He's surrounded by several people, most of them women. One is the woman from the party. I tell Jen he looks busy.

"I don't recognize most of them," she says. She leans in closer. "I sometimes wonder how many real friends he has. With all that money." She quickly downshifts into a smile when Paul looks up at us. "Go talk to him," she says.

"Warn me. Are any of them his ex-wife?"

"No," she says, "but I recognize a few ex-girlfriends."

Great, I think. A harem. Jen draws my attention to the woman standing to Paul's left. She has teased blond hair and heavy makeup and a body that looks as if it sees a lot of gym time. She was his first girlfriend, Jen confides, adding, "Paul says they're just friends now, but . . ." She drops the thought, but I can tell she wants me to finish it.

"Oh," I say, "a friend with benefits."

"Exactly," she says, "between relationships, I think."

As I approach them, Paul says something to the blonde. She turns to look at me and moves away as I enter their circle. Paul's eyes are glazed, as if he's in shock, or sedated. He leans toward me, and I instinctively reach out and touch his shoulder. He brushes my cheek with a kiss and says—whispers, really—"Thanks for coming. I was hoping you would." When I withdraw my hand, I recognize this as a moment that seals a connection.

Within days, he is at my door again. This time he's called ahead. He looks better than he did the evening of the memorial service. He's more relaxed, and he's dressed in what I'm guessing is his uniform: a pair of khakis, a tan, and his hiking boots. He asks me on what he calls a second date. "My father's memorial service was our first," he laughs. I smile, but the joke makes me uncomfortable, though I'm not sure why.

What I *am* sure about is that I'm nervous at the prospect of dating Paul. It's been more than a year since my last real relationship ended. I feel rusty, out of practice. Dating always makes me anxious. It's a horrid way to get to know someone. In general, I don't make many mistakes, and I'm tired of making mistakes in this area of my life. The pattern is clear: I pick men who are attracted by my strength and who then try to wring it out of me. It's not unlike my relationship with most of the men in my family. But that's only what makes the men I choose feel familiar. Something deeper explains my choices.

Paul asks if I'm free the following weekend. I tell him we'll have to wait until after the women's center benefit. He asks for details, then says that he'd like to buy tickets. I tell him that my mother is coming to town for the event, and that I'll be busy most of the evening, greeting people, giving a talk.

"That's okay," he says. "I'll bring a friend." He buys two top-price tickets, which include a champagne reception before the reading and a catered dinner on stage with Claire Bloom after.

When he hands me the check, I realize it's for twice the face value of the tickets. "This is too much," I say. "You've doubled it."

"I meant to," he replies. "It's for a good cause."

So a few days later, before we've even gone on an actual date, Paul meets my mother. He brings the same woman he took to Jen and Doug's party. I introduce them to friends, the president of the college, Claire Bloom, and, of course, Mom. It's strange: It almost feels as if we're involved already, but once again, I'm getting a strange vibe from his "friend."

Throughout the evening, I notice that Paul looks comfortable, making conversation with a number of people. At one point, when he is busy talking to my mother, his friend says to me, "I'm glad Paul's met you. It's such a hard time for him. It's nice to see him so excited about being with someone." I'm surprised by how relieved I am to know I misread her feelings.

"What did you think of him?" I ask my mother in the car on our way back to my apartment.

"He seems nice," she says, "but it's too soon to tell."

We talk about the performance and the food, and about the neighborhood we're driving through. It's one of Mom's favorite streets in this city, a wide avenue with historic mansions on both sides. While some of them are lovely, like the limestone Italianate villa that dominates its side of the street, many, such as the one just ahead of us—a mammoth, dark sandstone mansion built by a famous railroad baron at the turn of the last century—are just plain ugly.

"That one would make a better prison than a home," my mother muses.

* * *

Paul and I have our first official date the next Friday. The September evening holds the residual heat of summer. I choose a variation on the outfit I wore the first time we met. It's my way of not making the event seem too important. To the cobalt silk skirt and white silk shirt, I add a black silk sweater. As soon as the sun goes down, the air will remind us that in the Midwest mid-September is autumn, no matter what the calendar says.

He wears his loss like a well-earned scar. It feels surreal, the excitement of getting to know someone new in the midst of fresh grief. We stand on my balcony, leaning against the black metal railing, and he says, "I see why you love it here. You must feel like you live in an expensive tree house." He's got on a summer-weight wool sports jacket. As we walk out of my apartment, I brush my fingertips against his sleeve to assure myself that the fabric feels as lovely as it looks.

He's made reservations at a French restaurant. One of the owners is a famous cyclist, and it's on the latest hot list. It's a tough reservation, especially on a weekend night, but Paul's a regular, so getting a table and lingering for hours over our meal isn't a problem.

He doesn't drink. I check the backstory.

"Recovering alcoholic?" I ask, as casually as I can manage.

No, he tells me, he just doesn't like the way it tastes.

I can make a glass or two of wine last through dinner, but I'm used to men who can drink more in a day than most can manage in a week. I like that he isn't one of them. Over pistachio-crusted salmon and grilled summer vegetables, we pick up the threads of our earlier conversations. It hasn't taken me long to glean that Paul's not much of a reader, but, like me, he's a movie buff. As soon as we start ticking off the movies we've seen recently, it's clear that we have enough similarity and enough difference in our tastes to make it interesting. We both loved *The Fisher King.*

We split on *Thelma and Louise.* He liked *Sleeping with the Enemy.* It gave me nightmares. He's looking forward to seeing *The Prince of Tides.* I recommend that he read the novel first.

As the waiter arrives with the cake we've decided to share, I twirl the stem of my wineglass and steer our conversation in a more personal direction. While I'm secretly relieved that he has no family for me to meet, I want to know about them. I ask what his mother was like. "There's really nothing to tell," he says. It's clear he doesn't want to talk about her. All he offers is that when he was in his twenties, he found her in the kitchen, dead of a heart attack. She hadn't shown up for a golf game, and her friends asked Paul to check on her. A curious echo of his father's death.

The fact that he talks more about his father seems natural. It's a new loss. "My father always said he loved me," Paul says at one point, "but he didn't appear to like me very much. He belittled me every chance he got." I want more details, but it seems too soon to press. "If I do nothing else with my life," he says, "I want to be a better man than my father was."

Paul talks about being in therapy, so I assume more will come. He appears to be a man who believes in self-examination.

We nearly close the restaurant, still talking about films as we leave. I try to explain why I was more disturbed by *Sleeping with the Enemy* than *Silence of the Lambs.* "I know it sounds strange," I say, "because *Silence of the Lambs* clearly has more graphic violence. But in *Sleeping with the Enemy*, she's hunted by someone who's meant to love her, not by a stranger who's a psychopath. I find that much more disturbing," I say. He says he gets my point but doesn't see it that way.

As we leave the restaurant, the car's headlights flash on old oak trees scarred with bright yellow marks designating them for removal. Victims of a disease whose name I can't remember. I've written about it, but once I'm done with a subject I allow myself to retain only the essence of what I learned. Years earlier, as a stringer for *BusinessWeek*, I had to file stories about toxic waste, art investment, and luxury boat sales—all within a single week. By that Friday I decided that if I didn't let go of at least some of what I'd crammed into my head, I wouldn't have room for whatever came next. It takes too much energy to remember everything you learn. At least that's what I tell myself. What I recall, I recall clearly and in great detail. Conversations, for example: Some would say I can play back a conversation with an annoying level of accuracy.

I suppose if I were smarter, I wouldn't have to frame memory this way. I suspect *really* smart people don't forget anything they learn; they just keep on tucking stuff in their brain, day after day. I console myself with the knowledge that *really* smart people often have trouble navigating the daily course of life. I suppose all that stuff clogging their head makes it hard to see out, to deal with the rest of us.

Paul pulls the car over to the curb. We're in the old part of a suburb that borders the city. It's populated by medium houses on small lots. Colonial. Colonial. Mock Tudor. Bungalow. Craftsman. Bungalow. Ranch. Colonial. One block looks nearly identical to any other. My window is halfway down, letting in the sweet smell of burning leaves. The day's light has all but lost its power.

"May I kiss you?" he asks.

There it is again: asking for permission. I'm only half surprised, not by the idea, by the request. And, I think: Good plan. Get it over with. Then we don't have to worry about the if or when or where.

It's a careful kiss, almost chaste, as if he's afraid to give in to it entirely.

* * *

When he suggests we go to his house, I say yes, but don't intend to stay the night. I've heard about it from his aunt and uncle, and I expect something spectacular. I know that it was once his parents' home and that he bought and gutted it several years ago. The exterior is white and hints at warmth even though it looks as if it can't decide whether it's modern or traditional. Inside is something else entirely. I wonder what I'm missing. It appears to be waiting, for someone or something to make it look less perfect, less sterile. This house reminds me of what I don't like about southern California and the Arizona desert: It feels like a place without seasons.

Clearly, I don't see it as I'm meant to. In the living room, yards of expensive white tile serve as the backdrop for a pool table, a brown leather couch, a glass tabletop floating on burled wood, and a fireplace that looks as if it's seldom lit. The corners of the room are dark; there's hardly any light, except over the pool table. What appears to be a side solarium turns out to be a hot tub. The whole thing resembles nothing so much as a party room in an upscale condominium building.

We walk past a pristine kitchen and dining room, and up wide, thickly carpeted stairs, which give out onto a landing with a large window covered by a long panel of venetian blinds. As we turn onto the second flight of stairs, we're met by his father—or rather, a life-size acrylic portrait of his father—stretching across the wall. The thing must be at least four feet wide and six feet tall. He's meant to look friendly, all smiles and expensive blazer. But it's disturbing nonetheless, and not just because of the size. This man, dead or not, has eyes that frighten me. I remember how it felt to meet him, how I chose not to touch him, even with a gesture as detached as a handshake. It's as if I'm meeting him all over again.

"Is this new?" I ask, thinking that perhaps the picture has been hung in the wake of his father's death. I realize now that the upstairs is basically one large room and that this wall is in full view of the bed.

"No," Paul says. "It's been here for a while."

At the end of this combined bedroom and sitting room, which is carpeted in the same plush battleship gray as the stairs, is a fireplace sheathed in black marble. As with the one downstairs, it looks never to have been used. This place feels like a set for a David Mamet film, I think to myself. Paul gives me the tour. On the side of the room facing the backyard are windows and French doors leading to a wide balcony; at the end are two doors, one leading to a bathroom suite facing the street, the other to a bathroom suite facing the backyard. The one street-side is decorated in gray and black; the one facing the backyard, in gray and peach. His and hers. His has a large walk-in closet and a massive double shower. Hers has a walk-in closet, a shower, a tub, a wall full of drawers, and mirrors everywhere—on closet doors, the walls, the ceiling. It's meant to be gorgeous.

"This is like a shrine to vanity," I say. Oops. It's a flaw: letting words slip out before I think how they might sound. I can hear my therapist's voice: "You don't need to say *everything* you think. And when you *do* need to say something, remember: words that sound perfectly reasonable to you might sound a bit harsh to someone else."

Paul laughs. "It's what my ex-wife wanted," he says, not appearing in the least offended.

She must have liked to admire herself, I think, but do not say. I wonder how else I'm different from his ex. I can save that for Jen. She'll dish.

Back in the bedroom, my initial observation is confirmed: There's no way to be in that king-size bed and not fall under his father's gaze. But I do my best to ignore the father and concentrate on the son. This first night together, we lie, fully clothed, on his bed and continue the conversation we started over dinner. Already I think I could spend a great deal of time with this man. For hours, we're tentative in our touch, almost cautious. Several times I wonder what he's afraid of. That's how it feels to me, that he's afraid to start something.

My habit with a man has been to reveal my soul, and peek into his, before I'm ready for sex. Once that line is crossed, something changes for me, and I don't want to risk that feeling of deep connection before I'm ready. So what I mean to do with Paul is get to know him slowly. Midway through this first night with him, however, it occurs to me that strategy hasn't worked out so well in the past. At dawn, we make a definitive yet gentle move toward each other, with the care one takes with someone new.

Chapter 4

When Paul takes me home later that morning, I ask him to drive around to the back door. My apartment building is like a village. We have the nosy neighbors who watch everything just for the fun of it. The professionals who mind their own business. The gossips. The idiot. More than one, actually, if I were doing the counting.

Jen and Doug will want a report. I can tell from what they don't say that they're surprised I've agreed to go out with Paul. I put it down to their fear of awkwardness if things don't work out. I intend to be as closemouthed as possible. But Jen is good, the perfect mole. She cajoles the juicy bits out of you every single time.

The elevator in my high-rise opens directly across from Jen and Doug's apartment. I don't want the sound of the doors opening to announce my arrival, so I get off one floor below my own. As I walk up the stairs, I feel like a teenager sneaking home after curfew.

I've just closed my apartment door when the phone rings. It's Paul, calling from his car. He tells me he'd like to see me again, tonight, he says, if he didn't already have plans he can't change.

"I'm not free tonight anyway," I say. Actually I have no plans. So yes, technically I'm free, and I'm half eager. But my other half is wary. I haven't decided what I want to do about this man yet. I've been taught that when people seem too good to be true, they probably are. Paul's interesting, smart, attentive, and, if not classically good-looking, he has a style—a flair—I find intriguing. Yes, he seems too good to be true. But, I think, give him a chance. Wait and see.

He asks if I'm free the next day, and when I don't answer immediately, he adds, for a walk in the afternoon.

This sounds like a safe enough follow-up to our first date, so I say yes.

In almost all things, I'm a diligent sort of person. I've worked for myself for nearly twenty years, and, no matter what the work, on the job, I always come prepared. But when it's a matter of sexual attraction, when I meet a man who seems bright and gentle and not too sexist, I don't stop to prepare my heart. Wisely or not, I simply turn myself his way. It's a pattern I'm trying to break.

A history of disappointing relationships has led me to a clear conclusion: I don't want to deplete myself shoring up a man. I have seen close up what it's like to be trapped in love, have watched my mother and my grandmother devote their lives to men whose addictions wore everyone down. Even as a child I knew I didn't want this kind of life. I know what it's like to have home feel like the unsteadiest place on earth. In the end, things may have righted themselves for my parents, but I'm taking no chances.

Meeting Paul may have me reconsidering my decision to remain alone, but still I'm wary of once again choosing a man who looks like an adult and turns out to be a child. If only human mating had easy markers. Like the weather: red sky at night, sailor's delight; red sky in morning, sailor's warning—or shepherd's, if you live inland. Or when tapping maple trees: the higher the tap, the sweeter the sap. Or picking morels: "It's time when oak leaves are the size of squirrel's ears," a friend informs me one day on a walk. I want markers as clear as that for picking mates.

But life's not that simple, so I resolve to take this one slowly. I have plenty of opportunity over the next several weeks to add to my impression of Paul, for he calls every day, always more than once, and we see each other often. It's obvious the most pressing thing on his schedule is his father's murder investigation, but he makes plenty of room for an ordinary life, at least with me. We spend entire afternoons looking at art. At the Jenny Holzer exhibit at the city's modern art gallery, we're the only ones in the room. As her *Truisms* runs across the electronic board high on the white wall, we begin reading them aloud, like fencers exchanging thrusts.

"Being happy is more important that anything else," he reads.

"Every achievement requires a sacrifice," I parry, and, assuming we're not taking turns, I flick another into the room, "Ensure that your life stays in flux."

"Men are not monogamous by nature," he counters, and quickly follows with "Moderation kills the spirit."

Okay, I really want to challenge both of those, but I refrain. Instead, I ask, "Are you picking some of these out just to goad me?"

He reaches out and pulls me to him. "Yes. You're such an easy mark. I can tell by your face that you can hardly keep yourself from arguing every point."

"It's part of my charm," I say. Or not, I think, depending on one's taste.

It's a Thursday, and we've decided to take the day off. Well, I have anyway. Technically, he's got every day off. When we leave the gallery, we plan on seeing a movie, then going on to dinner.

One of his woman friends, an associate professor at a nearby university, recommended a Canadian film, *Strangers in Good Company*. It's about eight women, strangers, who get to know one another when their bus breaks down in the Canadian wilderness.

The movie lives up to its billing: entertaining and thought provoking. At dinner, back at the same restaurant where we had our first date, over the same pistachio-crusted salmon, we talk about how we're sometimes forced to face ourselves when we're caught in circumstances outside our normal lives.

"Like the women in the film," I say.

"Like me," he says, "with my father's murder."

I can't think what to say, so I reach across the table for his hand. I feel overwhelming sadness for him.

Our first come-to-Jesus meeting has to do with Rollerblading. And reading. At least that's how I frame the discussion. He loves Rollerblading. I love reading. I have no interest in even trying Rollerblading. My idea of exercise—other than walking and modest hiking—is something you do on a mat, like yoga and Pilates. The only time I don't mind sweat is during sex—just one more bodily fluid then. Paul, on the other hand, likes anything that promises burn. As for sweat, the more the better.

The third time I tell him I'm not interested, he switches tactics, from cajoling to persistence, asking how I can say I don't like it if I've never tried. "I'm a really good teacher," he says. He assures me he's patient and kind. "I've taught lots of people."

Good point, I think, knowing it's *one* way to put the subject to rest for good. I agree to try. It's a perfect autumn day. I rent skates, and we set off along a lake path paved especially for cycling and Rollerblading. Two hours later I can see he's right. He's a good teacher. But I was right, too. I hate Rollerblading. I'm terrible at it. I grew up ice-skating, and I can't seem to find the rhythm of blades on the ground. I can see this bothers him.

A few days later he asks if I'll try again. "Look," I say, after I refuse, "I wish you were a reader. Most of my life revolves around reading, and I wish you loved books as much as I do. But you don't, and in the long run, I care more about being with someone who's honest and kind and generous and smart." I tell him if he doesn't feel the same way about being with someone who's not as athletic as he is, now's the time to discuss it, because neither of us is going to change substantially.

Of course, other things are more important, he pouts, as if I've just criticized his first child. "Especially when you put it that way," he adds. And we both start to laugh.

"So we're clear about that?" I say, "because I'm not going to turn into an athlete just to keep you around. And I don't want you to feel as if you're settling."

"Wanna go to a movie?" he says, reaching for the newspaper.

A few times a week we meet for a walk, sometimes around one of the city lakes, sometimes along the Mississippi. Some Saturdays, when the weather's bad, we go to two or three movies in a row. If we've been together Saturday night and neither of us has plans for Sunday, we pick up a *New York Times* and drive to a coffeehouse, where we sit for hours, trading sections, comments, thoughts.

We fill our calendars with dates for concerts, dance performances, and literary events. Planning ahead where we live often involves baseball, basketball, football, and hockey. Being a loyal sister to athletic brothers gave me more than my fill of that. I'm relieved Paul's not interested in spectator sports either. One night we go to see a movie and, confused to find we're the only ones there, head back to the box office to ask what's going on.

"The World Series," says the kid in the box. He can't believe we don't know that our state's team has made it in this year.

I love that Paul's as clueless about such things as I am. He's more interested in private competition, and unexpected hobbies. He wants to start collecting bonsai. I've always admired the carefully clipped plants in a vague sort of way, but I find them artificial, too contained. I prefer beauty that's haphazard; I'll go to an exhibit of bonsai, but I don't want to live with one. But Paul's become enamored of them, and we visit specialty nurseries in search of a tree for his house. One afternoon, while he's in conversation with the owner of a particularly well-stocked nursery, I wander over to another section of the nursery to look at pots. I overhear a woman behind the counter talking to another customer. Something in the tone of her voice makes me stop to listen.

"No," she is saying in a high-pitched voice. "It's not too late to cover those plants with that black dirt we bought last spring. Like I told my husband, we missed it last year doesn't mean we miss it this year. Not if we're ready." She sounds a bit manic. I turn to match the face with the voice, and she looks up and smiles at me. She's thin but strong, with the kind of wiry muscles you get not from working out but from plain hard work, a lifetime of it. Her straight brown hair is tucked behind ears that look too big for the rest of her. Her eyes look scared, as if she might be punished if she's wrong about what she's saying.

It's a long-held joke with my family and close friends that when I leave a place I may not be able to recall the most glaring physical details, but I can gauge the emotional tenor of a scene precisely. A younger brother once confided that he doesn't even like to be in the same room with me when he's not ready to face something hard. "Not that you'd say anything unkind," he says, "but I can tell by looking at your eyes you've guessed what I'm avoiding, and it makes me nervous."

That's how I feel now. I won't remember the exact size and shape of the pots or the bonsai, but I'll never forget this woman, and her uncertain voice, and her anxious eyes. I try to focus on what she's saying, but I can't get past wondering why she's frightened, and I know I'm right about this. Her husband? That's what I think, though I don't know why. She turns back to the customer she's helping, a middle-aged woman dressed in matching camel jacket and trousers, who's actually beginning to look a little nervous herself. I can tell she's eager to leave.

"Like I told him," the woman behind the counter continues as she hands the customer her package, "it's getting colder, but we've still got time. Today or tomorrow we've got to do it, though, or we'll miss it again."

I'm not a gardener, so I can't even pretend to understand what she's talking about. I have no idea when "too late" happens for plants in the fall. I don't have much interest in actual gardening—too much dirt and sweat involved—but I do sometimes wish I could turn my antennae away from people's feelings toward something less complicated, like when it's time for black dirt.

As usual, we walk out of the nursery empty-handed. When Paul wants something, he's tireless in his search. He won't stop hunting until he has exactly what he wants.

Chapter 5

Whenever Paul and I part, whether we've been together for a few hours or a few days, his good-bye kiss is soft, like twilight, filled with gentle promise of things to come. He tells me nearly every time I see him that he can't believe his fortune meeting me. Especially now, he says. "You've been so patient and kind. A lot of women wouldn't have put up with me." Then he adds that he's glad we didn't meet sooner.

I ask him why.

"You wouldn't have liked me if you'd met me when I was younger."

He's piqued my curiosity. "Why?"

"I wasn't grown up enough for you."

In fact, Paul is reluctant to talk much about himself. Whenever I ask about his ex-wife or girlfriends, he deflects the questions. He shares little about his family, and what he does offer makes me sad. Paul's well off if you define the word only in terms of money. An only child, he inherited a fortune from his mother when she died. In fact, both Paul and his father benefited from her death, assuming ownership of companies requiring no management on their part. His father's death left Paul even wealthier than before.

Now he owns two companies that virtually run themselves, plus nearly all his father's other assets.

Paul tells me enough to make one thing clear: he was deprived of what I take for granted—a house full of love, no matter how imperfect. Nothing he tells me suggests his mother and father ever loved each other—or him, for that matter. Not as I understand love anyway. Spoiling a child is a particular form of neglect in my book.

If you're an even remotely optimistic sort of person, and you have a difficult relationship with one or both of your parents, you never quit hoping for that single moment when your parent looks at you and says, meaning every word: "I'm sorry. I should have done it better. I love you. Please forgive me." And then you do your part: you forgive—graciously, authentically—and take responsibility for your own failures.

That moment has passed for Paul with both his parents. This strikes me as the true grief he'll never get over. So I do my best to offer him what I can—affection, empathy, ordinary days spent doing the things ordinary people do together: taking walks, visiting friends, running errands. It occurs to me the most valuable thing I can share with Paul is something he needs more than anything else: a sense of connection, and of family, which is something I have plenty of, even if far from perfect.

Imagine a midwestern version of *Bonanza:* a savvy father, a wise mother instead of a Chinese cook, five sons instead of three. Then add in three daughters and a steady stream of alcohol running through the family, and people doing and saying things they wouldn't dream of sober. After factoring in the Catholic code and the Irish temper, what's left is the current that never ceases—parents who love us. That's my family. In a word: complicated.

* * *

A blizzard. On Halloween. It can happen. I don't expect to see Paul for several days; he's going to Albuquerque to pack up his father's house. When the heavy snow starts, Paul calls. "I'm not going to be able to fly out today. Let's get snowed in together. I can be there in half an hour."

I'm glad for the chance to abandon my one-room apartment for a house if we're going to be snowed in. On the way to Paul's, we stop for pizza, popcorn, and videos, and fresh-baked scones for the next morning. Like kids on a snow day, we walk in the evening fog of blustering snow. That night, after we shower together, we watch movies, wrapped around each other as we fall asleep.

We stay in bed the next morning long after we wake. He tells me I'll wear him out. "Tell me something I believe," I say, and turn his way again. Not since Jim have I been with a man so eager to please me sexually, though I'm bothered by how clinical Paul can be. With Jim, lovemaking felt instinctual. Besides, the one time I don't like a lot of conversation is during sex, and Paul is full of questions. I tell him not to worry so much, but he insists he wants to make me happy. Then shut up is what I want to say, but don't. And save all the chatter for the times I want you to talk and you don't. I want to say that, too.

He doesn't seem able to bring a sense of humor to bed with him. At times he appears confident. At other times, he seems fragile, or absent. I chalk it up to the trauma in his life.

By late in the afternoon planes are flying again, so Paul leaves for Albuquerque. He calls each night but says nothing about his meetings with the police. Instead, he tells me what he's finding at his dad's house. Shall he take the lead crystal, he wonders aloud. Not waiting for an answer, he rushes on about a great set of Calphalon pots. "Maybe I'll sign up for some cooking classes after this is all over." It's as if appropriating some of his dad's things offers him hope of a fresh start.

* * *

In the background of our time together runs a sinister sound track, an incessant murmur of details about the murder. Even though Paul doesn't talk much about the specifics, he's consumed by the investigation, talking to the police several times a day. Finally, on November 4, exactly two months after the murder, the police make an arrest.

It's Jen who fills me in on the specifics: they've arrested a man in his twenties. Roger knew him. Her brother had a habit of finding young men who were either in prison or recently released, and offering them money, the police inform her. And then, when the young men came to depend on him, Roger would withdraw his resources.

Apparently, the suspect wanted another loan, and Roger turned him down. Then one night, he appeared at Roger's house. Car trouble, he claimed. Within minutes, he'd taken a gun out of his jacket pocket and forced Roger into his bedroom, told him to undress, and bound him. He used Roger's own belt to strangle him. And videotaped it all.

When I hear these details, I am hit with a detached sense of horror for Paul's father, but mostly I feel nothing for the dead man and everything for the son. And for Jen. She's lost her brother in the most profound way: she realizes now that she hardly knew him at all.

More than anyone, my mother encourages my professional career, so it's fitting she's come to hear my paper at the Midwest Modern Language Association conference in Chicago. As I stand at the podium, I'm relieved to see the room so full. It gives me confidence to state my case: "In the same room in which Anne wrote,

Charlotte later sat at her desk and deleted stanzas from Anne's poems, softened her language, cut and rewrote her prose, and—not content with that—refused to grant publishers the rights to reprint Anne's two novels, *The Tenant of Wildfell Hall* and *Agnes Grey,* after her death.

"When I think of Charlotte Brontë as having edited Anne Brontë, I think of the phrase 'managing the news,' for Charlotte did not just edit Anne's work, she edited her life as well. The two are intricately connected. I believe that by editing the facts of Anne's life, Anne's life and work were for decades viewed with a certain prejudice, a prejudice that made way for condescension by future editors, publishers, and scholars."

Forty-five minutes later, my presentation is met with what I expected: both praise and criticism. Loyal Charlotte scholars don't like to attribute anything negative to their literary hero. I'm actually more surprised by how much praise I receive. As soon as I've finished speaking and ask for questions, one woman stands and thanks me for having the courage to reevaluate this family's history.

Now Mom and I have the weekend to enjoy the city: museums and bookstores, lunch at Marshall Field's, and tea overlooking the Water Tower. Paul has angled for an invitation, but I wasn't even tempted to change my plans with my mother. He calls the hotel several times and leaves messages. They all end with "I miss you."

I'm both embarrassed and delighted by his constant attention, which is unlike anything I've known before. No matter where I am, he calls several times throughout the day. Paul has no profession to speak of, just a list of activities he creates to fill a day—lunches with friends, movies, pro bono photography, and endless phone conversations. At times, it feels as if his waves of attention might tip my balanced life. But I'm determined to find a way to avoid getting swept away by his attentions.

* * *

Shortly after I get home from the conference, my sister Liz and her young children come for a visit. She lives in the same town we grew up in, a few hours drive from my home. The first night she's there, after the children are tucked into their sleeping bags at the foot of my bed, she notes how persistent Paul is and how resistant I seem to go out with him as often as he'd like. Since she arrived a few hours ago, he's already phoned twice. She asks how often he calls.

"A lot." I'm not eager to do a count.

"Every day?" she asks, and when I nod yes, she presses: once, twice, several times?

Most days five or six times, I admit. "He says he likes to know where I am, what I'm doing. He wishes I had a cell phone so he could reach me wherever I am." I try to laugh it off when I tell my sister what I've told him: if I get a cell phone it won't be because he wants to reach me any time of the day or night.

My sister is more than a decade younger than I, and both more and less experienced. Married to her childhood sweetheart, she knows much more about marriage than I do, but much less about dating. We're close, but only as long as I keep things relatively light. She doesn't like to go deep if she can help it, which is, of course, where I prefer to live. I find it difficult to articulate my sense of unease, but I'm clear about one thing: it isn't just his money. Finding balance in the midst of his wealth and his constant attention would be tough enough, but that pales in the glare of the heightened drama of living through a murder investigation. And there's something else: I'm having trouble putting the pieces of Paul's life into a coherent whole. I'm used to seeing things clearly. With Paul, it's like looking through a scrim.

When I tell Liz I'm trying to take it more slowly than Paul would like, she suggests we invite him for dinner. So I do. While she and I prepare roasted chicken and vegetables, and the children's favorite—apple crisp—in my galley kitchen, he entertains her toddlers. We watch him with her blond, curly-haired daughter and son. I know she's ticking off items on her internal checklist, as am I. He plays Candyland with them. He reads *Make Way for Ducklings*, throwing in the occasional quacking sound. They climb all over him, and he sends them into the hysterical hiccupping giggles only kids are capable of. He appears to be a natural with children.

At dinner, Paul asks Liz about her teaching. My sister is known for her ability to keep abused kids from getting lost in the system. We all share an interest in working for children. I'm on the board of a residential center for abused adolescents, and Paul is involved at a school for children with learning disabilities.

After Liz puts the children to bed in my room, she and Paul and I talk for hours. After he leaves, she tells me how much she likes him. "He's fun," she says, "and smart. And it's obvious he's crazy about you." She laughs, "Don't hold his money against him."

I know she could be right. I do harbor a prejudice against a certain type of very rich. Too many wealthy people I know seem to view the world in a slanted way: they think the world owes them. They don't understand—and sometimes it's a matter of forgetting—the choices most people make every minute of every day. I want nothing to do with people who have resources but no heart, people who are charitable but not kind.

That night, I resolve to drop my resistance and take Paul as I find him, not as I'm afraid he might be. And I find him generous, always looking for places to give money where it will make a difference—a local homeless shelter, a new program for victim advocates. I don't know all the details, but I know he's helping support the three young children of a friend who's in and out of treatment for drug addiction.

Besides, I like being with someone who notices my interests. In between our scheduled dates, he calls often, sometimes asking if he can just drop by. Just for a moment, he usually says. When I open the door, he stands with a smile and a package—sometimes something playful, a kaleidoscope; sometimes something serious, a book. One day I mention a Charlotte Perkins Gilman book that's difficult to find. Within days, he delivers a copy of *The Yellow Wallpaper* to my door.

And just often enough he tells me he thinks I'm beautiful, with a style he admires, in looks and in life. I'm not immune to flattery. For all my apparent strength and confidence, I'm desperate for such compliments, especially the ones that still voices from my past: *Slut. Coldhearted bitch*—words my father drilled into me during his drunken rages.

We're sitting in a theater waiting for the movie, *Frankie and Johnny*, to start. Paul is telling me he envies the sense of certainty I seem to possess. He says he's never felt free to go after his dreams.

"If you could do anything, anything at all, regardless of the money, the education, the training, what would you do?" I ask.

"I'd be a doctor," he says. "I've always wanted to be a doctor."

I'm shocked. He's never even hinted at this. I ask what's kept him from going to med school.

He tells me he didn't think he was smart enough. "My dad made sure of that," he adds. He works his mouth and jaw, as if grinding his next words into shape. "Besides," he says in a more subdued voice, "I've got learning disabilities."

I'd guessed as much but hadn't asked. I figured he'd tell me when he was ready. "You've got all the resources and time anyone could want," I say. I encourage him at least to try. "Apply to med school. Get tutoring if you need it."

"It's too late," he says.

"It's *not*," I say. "Think of all the people our age who realize going for their dream means ignoring spouses or neglecting children. You can do it and no one will suffer." I look at him. "If you don't try, you'll always wonder."

I let that sit. "Besides, Paul," I say, and even as the words leave me I wonder how I dare, "he's dead. Your father's dead. His voice can die now, too."

He's still looking straight ahead. When I can't see his eyes, I can't read him. The light dims, and the movie score begins. As the first title scrolls across the screen Paul laces the fingers of his hand through mine. "Thanks," he says, so softly I almost miss his next words. "No one's ever encouraged me like this."

We turn to the screen to watch two damaged souls, saving each other.

Chapter 6

The small bronze urn with Roger's ashes sits on the shelf near Paul's bed for several weeks. I've been eager for him to bury them, and finally he does. He's invited only his closest friends—and family, of course—to the graveside ceremony and dinner at his house afterward. I've introduced him to a friend who owns a Greek restaurant, and Paul has him cater the dinner. The *dolmathes*, *spanakopita*, and *shish kebabs* remind the family of the time Roger took all of them—Paul, Jen, Doug, their children and spouses—for a cruise around the Greek islands. They make it sound nothing but wonderful, but I'd heard about some of Roger's verbal attacks, belittling Jen and Doug especially. Roger needed people to feel unsure of themselves, and as usual, it sounded as if he'd succeeded.

The next night as we cross a parking lot after an errand, Paul pulls me to him. "I've been waiting to tell you something." He looks like a child, eager to share a surprise. I think he's planned a trip. He's talked of our going to New York together; perhaps that's it. He holds me at arm's length and looks at me with an intensity I haven't seen in him before. "I love you," he says. "I love you like I've never loved anyone before."

I'm not ready for this. My face must say so. We've known each other for only four months. I wonder if this is a reaction to the renewed grief I saw on his face the day before.

He rushes on. "I don't want to scare you away, but I couldn't wait any longer to tell you." He pulls me close. I can no longer see his face. "I love you," he says again. "I love you."

It's nearly eight o'clock. Cold air gathers in foggy halos around the high lights in the parking lot. I focus on one. Paul's so much bigger than I am, when he hugs me like this I'm lost. "I don't know what to say," I murmur into his shoulder. I pull back, thinking I should offer something in return. But I don't know how I feel yet.

"You don't have to say anything," he says. "I just want you to know how much I love you."

I'm touched, but I want him to stop talking. "I'm glad," I say, "but it's too soon for me to know how I feel." Even as my next words come out, I sense that I'm only making the situation worse. "But thanks for telling me." I watch the eagerness drain from his face. "I think I love you, too," I quickly add, but he can see I'm just saying it to make him feel better.

"It doesn't matter," he says. "All I know is that I love you, and I want you to know. I feel as if I've known you my whole life, and I want to know you forever."

I don't feel as if I've known *him* forever. He's hard to know. I can't tell how much of the blurriness surrounding him is due to the turmoil of his father's murder. I'm beginning to grasp the depth of his loneliness. What I first saw as charming vulnerability now worries me. I've discovered he's not really close to his relatives. Jen tells me they're seeing more of him now that we're dating than they have since he was young. "He's always kept his distance," she says. I'm happy to bring them closer, but it also makes me feel uncomfortable. It might mean bigger fallout if the relationship doesn't last.

All this is making me queasy—excited, but queasy, too. I don't know if what I feel is love. Paul doesn't seem threatened by my strength, by my work, but I've seen before how that can change. Yet, in the light of his pain, how can I not respond in a loving way? I *do* feel compassion. It seems the least I can offer; it seems heartless not to.

But still, I'm afraid. This feels different, unlike anything I've known before.

The first time Paul meets the rest of my family is at an uncle's funeral. We get up before dawn the morning of the service and drive nearly two hundred miles to the town I grew up in. Because we've had such early snow, and so much of it, the highway is lined with huge snowbanks. Rather than lifting the landscape, the tunnel effect renders predawn even drearier, making everything seem an effort.

After the funeral mass, we all huddle together at the cemetery, hoping to keep warm. Paul draws me close, giving notice to my relatives of his feelings. We stay for the lunch, which is held at one of those places found only in towns of a certain size, the all-purpose cavernous room furnished with a heavily patterned carpet, industrial weight to hide all stains, and faux leather furniture, sturdy enough to accommodate even the heaviest guests.

In large, boisterous families like mine, siblings often throw out challenges for newcomers, who are, in turn, meant to prove their worth. To be taken in, you must give back as good as you get. Paul doesn't miss a shot as they lob playful comments his way. Instead of being impressed by his wealth and life of ease, one brother teases him about how tough life is when you don't have to work. It must be a burden, he says, to have to fill all that time.

The joke is meant to replace the usual easy exchange about what people do for a living.

Watching Paul laugh and banter back with good humor, it strikes me that he seems to fit into my brothers' game more easily than I do. In my family, learning the fine line between wit and sarcasm hasn't always been an enjoyable lesson. I remember years ago looking up the word *sarcasm*. From the Greek: to rend flesh. That's how it feels when wit turns mean. Someone gets hurt. I also note, not for the first time, that my siblings are more comfortable with me when I'm part of a couple. Perhaps a mate always serves as a buffer in a family. If so, the cushion works both ways. Before we leave for the day, one by one, my brothers tell me how much they like Paul.

My mother's vote comes in the form of food. As we walk to my parents' car to say good-bye, Mom reaches in and hands Paul a container of her fudge bars. "I hear you love chocolate," she says. He sends a half-shy, half-cocky smile her way and kisses her on the cheek. We all know how to interpret her gift. Mom has stamped him: Approved.

I'm too busy filtering everyone else's feelings to be much bothered by my father's lack of warmth toward Paul. Since his stroke twelve years earlier, his manner is more reserved. I put it down to that.

Paul goes to Hawaii for Christmas and New Year's. "I need to be alone," he says. I assume he wants to get away from everything that reminds him of his father's death. Much of his time is spent sorting out his father's estate, and listening to the police and attorneys talk about the pending murder trial. His dining room table is covered with file folders and sheets of yellow legal paper. And lots of envelopes from attorneys.

Even if he wanted me to go with him, I don't want to go. I'm not crazy about beach vacations. I'm one of those people who burn rather than tan. And I'm not ready to invite him to our family holiday, which I'm spending at my sister's home, a few miles from where my parents live.

The first several days of his two-week trip Paul calls every day, usually more than once. I'm busy with my family, sledding or walking or reading books in front of the fire, so sometimes I miss the call. When I talk to him, he sounds as if he's busy, too—hiking, swimming, scuba diving. He says he misses me.

A few days before he's scheduled to fly home, I call his hotel to confirm his arrival time at the airport. The hotel receptionist tells me there's a "Do not disturb" message posted on the computer. I leave a message. No callback that day. I call the next day and the day after that. No callback those days either. The message is still in place, according to the operator.

I drive away from my sister's house on a bright, cold January day not knowing what to think. Actually, I know what to think. I just don't want it to be true. I suspect Paul's with someone. But I can't know for sure, at least now.

As I'm unpacking, Paul calls from the Honolulu airport and asks if I'm still planning on picking him up. I tell him yes. He sounds rushed, or preoccupied, so I ask him if he's running for a connecting flight.

No, he says.

I ask him why I couldn't get through to him at the hotel.

"Oh, I took a nap one day," he says, "and I forgot to have them take off the 'Do not disturb' request."

"And the reason you didn't return my calls, was . . . what exactly?"

He tells me he didn't get them until he checked out.

I want to believe him, but I don't. "Were you with someone?" I ask.

"No," he replies. Even across the distance I hear his voice turn cool. "Why would you even *think* that?"

"I'm sorry," I say, "but I wondered. It seemed strange."

All the way to the airport, I'm ashamed of myself for being suspicious, but when I meet him at the gate, he seems nervous. He rushes me through the terminal. "I can't believe you think I was with someone just because I didn't answer my phone," he says as soon as we get into the car.

I apologize. It's another sunny day, unseasonably warm for this time of year. We drive to the arboretum for a walk. His mood keeps shifting—playful to moody, affectionate to aloof—as if he can't settle in a feeling.

"I can't believe you don't trust me," he says.

I apologize. Again. I'm confused and hurt. This isn't the greeting I expected from someone who professes to love me. When I drop him off at his house, he asks me in.

"No, I don't think so," I say. "I'm tired."

He puts his arms around me and kisses me. Hard. Harder than ever before. I wonder if he notices he's doing all the kissing. I'm just sitting there, waiting for him to get out of my car.

Chapter 7

Heart beating more than a thousand times a minute—it's true, hard as it is to picture—a hummingbird can mistake a speck of red for nectar. I've seen one dart toward a flower in a window vase and be stopped cold by a flat pane of glass. Startled, misled, confused. That's how I feel. I make excuses to avoid Paul when he calls the next few days. I've apologized, but it doesn't mean I'm convinced he's telling me the truth. In fact, I'm certain he's not. Finally, late on the third day, I give in and invite him for dinner the next night. A few hours before he's due to arrive, he phones to tell me he needs to talk to me about something. I ask if it can wait until he comes over.

When he tells me, he says, I won't want him to stay for dinner.

"Want to tell me now?"

No, he says. In person.

I explain that I'm on a deadline. I tell him to come when we planned. "We'll talk," I say. "Then we'll see about dinner." I try to make it sound like a joke. I'm anxious, but relieved, too. I have little tolerance for mysterious tension. I put a chicken and root vegetables into the oven to roast, but I suspect that I'll be eating them alone.

The moment Paul walks in the door, before he even takes off his black leather jacket, he pulls me to him and hugs me hard. He lets out a huge sigh. Then he breathes deeply, in and out, still holding me tight. He rests his head on my shoulder. I know this is serious.

"When I went to Hawaii, I was alone," he starts, and then tells me he was with someone at the end of the trip.

I try to push him away. "Get out of here," I say. "Do you know how lousy you made me feel for suspecting you?" He doesn't relax his hold on me. I try to recall how many times I've apologized. I mostly recall his indignation. "Get out of here," I repeat.

"Let me explain," he says.

"You asshole!" I push against him even harder. Then I stand still. His hands are gripping my arms. "And let go of me. Let go of me . . . *right now*." There's no mistaking when I'm really angry: my voice flattens, and I lengthen each syllable.

He drops his hands and rushes into his explanation. "Before I left on my trip, a friend told me she was going to be in Hawaii at the same time, so we planned to spend time together."

I don't believe this "coincidence," but I'm so angry I don't interrupt.

"She flew out after me," he continues, "and we flew home together."

"A friend?" I ask. "Or someone you've been dating?" Now I know my instinct wasn't wrong. "Don't bother answering," I say, and point out that if she'd been a friend, he would have told me about her before and introduced me at the airport.

I was right to be suspicious. He says he started dating her in November because he wanted to make sure I'm the one for him. "But

I don't love her," he says. "She's not important to me. I love *you*."

I walk to the door. "Get out." I look him straight in the eyes. "And don't even think about calling me."

But, of course, he does. He calls later that night, several times. He calls several times the next day, and the day after that. I let all the calls go to the answering machine. His messages follow an uncomplicated script: I've been a fool. I love you. I've never loved anyone like this before. *Please*. Forgive me. Take me back. You won't be sorry. I'll prove how much I love you. Sometimes he calls from downstairs, from the security phone. Then his messages simply say, "I'm downstairs. *Please* let me in."

I don't answer the phone, and I don't return his calls. I avoid talking to Jen and Doug, but when I see them in the hall, I can tell they know something's going on. Paul must have talked to them. On the third day, his uncle buzzes him into the building, and Paul appears at my door. I open it just as far as the fastened chain allows.

"Please. Forgive me," he says, pressing into the small space between the doorjamb and the chained door. He's in his winter uniform: khakis, a cotton turtleneck, and his heavy leather jacket. I catch a whiff of the leather and of soap; I can tell he's just showered and shaved. I used to like the way he smelled. Now I find it repellent.

He stretches his hand toward me, but I back away. "I love you," he says. "Please give me another chance."

When I couldn't see his face, I resisted him, but I've been trained to take pity on men who look as pathetic and sorrowful as he does now. I give in. I open the door and set terms: regular counseling—with me *and* on his own. No sex—with me "or anyone else for that matter," I say to him, "if you want me to believe you love me."

And no promises on my part. I say that, too.

Chapter 8

January remains unseasonably warm. In the midst of my confusion, I see this as a gift—from no one in particular, as I don't believe in a deity who parcels out good weather as an antidote to emotional pain. But that said, had I woken to the biting cold of below-zero temperatures day after day, it would have been harder to face my mixed feelings about Paul.

In the joint sessions with our therapists, Paul unveils what appears to be a confused and penitent heart. The verdict is unanimous: his behavior, hurtful and contrary to his words, is his way of reacting to his father's murder. For a man, a father's death is always traumatic, says his therapist. Life changing even, he adds. In Paul's case, it's complicated by the fact that his father was murdered in a horrible way, and his secret sexual life was exposed in public. Both agree on that, too. "It's natural you're not yourself," his male therapist says time and again.

I work at forgiving Paul, and trusting him again, a process that feels both foolish and generous at the same time. Mostly, I wait for the trauma of his father's death to release its hold on him. But in fact, Paul becomes even more obsessed with the details that have come out about the case.

In the kitchen of his house is a built-in desk. Above it are open shelves and cupboards, and small cubbyholes, each a few inches wide and several inches high. Each time he enters the house, Paul heads directly for the desk. Into one goes his outside stuff—money, calendar, car keys, glasses. Any receipts or notes he's accumulated throughout the day he files in the built-in file drawers in the dining room. Paul's an extremely organized sort of person.

Another cubbyhole contains a small photograph. Not of his dead parents, his friends, or me, but of the man accused of murdering his father. If it's possible to look dangerous and listless at the same time, this man manages the pose. Sometimes Paul leans against the kitchen desk and studies the photo. The first time I see him do this, I ask him who the man is. "The guy who killed my father," he says, with no emotion in his voice. That's all. "The guy who killed my father."

The first time, I assume he's just received it in the mail. The next time I see him with the photo I ask why he has it. "I asked the police for a copy of his mug shot. I don't want to forget what he looks like," he tells me. It's become a ritual, and it seems to give him pleasure. I don't understand this.

One afternoon, when he picks me up at my apartment, he's on the phone. "How old is she?" I hear him say. He listens and says, "Look into some schools, and let me know what you find out." When he says good-bye, he adds, "We'll talk tomorrow."

"Who was that?" I ask. He seems to have wanted me to overhear the call, so I don't feel shy about asking.

"It's the woman who turned my father's killer in," he says, and explains that she's the man's ex-girlfriend.

"And why are you talking to her?" I ask.

He says he wants to help pay for her daughter's education.

"Do the police know about this?" I ask.

That's how he got her number, he tells me. "I just want to help her. Besides, this is none of their business."

"I suppose it's none of mine either, Paul, but I'm not sure it's such a good idea. For all you know, she's still in contact with her ex. Couldn't it compromise the case?" That's just one of many complications I can think of.

"Look, I want to help her. I'm grateful she turned him in. And," he says, "I don't want to talk about this anymore. You're overreacting."

Perhaps I am. After all, he's used to sharing his wealth like this. Maybe that's what I'd do, too, if I had more money than I knew what to do with. I feel ashamed for being critical of his generosity. And yet, everything about this seems abnormal to me. I don't know how to hold my concerns steady in the face of such turbulence. First there were suspects to pursue. Now, a trial to plan and attend. Further down the road will come the sentencing. It feels as if this will never end.

One evening, we go to my cousin Maggie's house, for a New Year's drink, although it's nearly February. She and her husband Greg are eager to get to know Paul, whom they met only briefly at my uncle's funeral. The warm spell has given way to the kind of winter nights that make you wonder why you live in such a place. The temperature is well below zero, and the wind chill is lower yet.

Paul and I don't stay at my cousin's long because we are going to a play. On the way to the theater, we pass a woman standing alongside her car. In this weather it can be only one thing. Of course we stop to help. I roll down my window. "What can we do to help?" I ask, then add, "Come in and warm up."

When she gets in the car, I can see she's a little younger than we are, and attractive. We take her to a nearby service station and then wait until her car is running. When she thanks us, Paul says, "Why don't you come to the theater with us? It'll be easy to get another ticket on such a cold night."

This isn't the first time Paul's done something like this. One afternoon in a movie line, he started a conversation with the woman standing alone in front of us. Just before we got to the ticket box, he invited her to join us for dinner afterward. When he makes such overtures to women we don't know, I'm so stunned I just watch it happen, and I'm grateful they're decent enough to say no. But this time I'm angry. At first, the woman declines, but Paul insists. I don't chime in. I can see she's tempted to give in to Paul's persistent charm and say yes. Luckily, she pays more attention to my face than Paul's as she determines her response.

As we drive away, I tell Paul how uncomfortable, and angry, this makes me. "You keep telling me how much you love me," I say, "but things like this don't make me want to love *you*." Normally I'd keep such a thought to myself, but I've made it clear this is a trial period for him.

He acts surprised, and says he doesn't understand. "I'm just being friendly. I like to meet new people."

"New *women*," I say.

He assures me he'll stop it if I don't like it, but his tone is tinged with a suggestion that this is my problem, not his.

I ignore the implication and say, "Okay. Stop it. I *don't* like it."

We're late for the play, not a very good one as it turns out. I find myself wishing I'd stayed home, alone, with a good book, or at my cousin's for dinner, without him.

In our therapy sessions, when I press him to talk about his pattern of being overly familiar with women we meet when we're out, he once again blames his behavior on his father's murder.

"I'm not myself," he says, then turns to me. "I'm sorry I'm hurting you. I don't mean to."

I'm not entirely convinced that there's a connection, but I let it go. After all, our therapists aren't disputing the link. I want to understand this man who's full of near-silent pain. I try to get him to talk about his childhood, but he works his jaw in anger every time I ask questions about his mother or father. He tells me so little. I want to believe things will get better if he'll talk things out. I also want to believe it's just a matter of time.

One night on the phone, Paul tells me he wants us to do something physical together. He asks if I have any ideas.

Judo, I tell him. I've always wanted to try judo. It's something neither of us has done before, and Paul is excited about our learning something new together. The next day I call around to studios, and within a few days we've bought our judo *gi* and white belts, and we've signed up for a class together. As we train, I feel stronger, more confident physically. Not an athlete, I'm surprised I take to it so well. Paul, on the other hand, who's much more physical than I, seems impatient with the discipline, even though he's no better than most of the people in our class.

In class, as we learn throws and hold-down techniques, the instructor makes sure we work with like-size students. One night, Paul and I arrive at class a few minutes late. Since everyone is already paired up, we practice together. As we get into position, the instructor reminds Paul that, because I'm a lot smaller, he doesn't have to use the same power to throw me as he would someone his own size.

Paul and I bow and grab each other's sleeves. He pivots his hip, and over I go. I land just as I'm meant to and quickly get up. Again, Paul leans into me: another clean throw and a smooth landing. The next time, I manage to move more quickly than Paul, and he's the one on the ground.

"Nice try," he says as he gets up.

"Better than a try," I say, smiling. "I *landed* you." I'm proud of having thrown someone so much bigger and stronger.

We get into position again, and Paul throws me hard, knocking the wind out of me. Then he helps me up. I assume he's made a mistake, so I try to make a joke of it, reminding him that the word *judo* is Japanese for "the gentle way."

As soon as we get into position, he does it again. My body slaps the mat like a thunder clap. "You don't have to throw me so hard," I manage to say as soon as I catch my breath. The instructor is watching us. "Not so hard," he says to Paul.

But, in fact, Paul throws me just as hard again the next time. When I pull myself up from the mat this time, I tell him I'll wait until it's time to switch partners. As we drive to my apartment after class, I ask Paul if he's angry with me.

He asks me why I would think that.

"The way you threw me at the end," I say. "The first time I thought you just hadn't judged my weight properly. But after that, I wondered."

"I don't get angry," he says.

"Yes, you do," I respond.

"When have I ever spoken angrily to you?" he says.

Well, right now, I want to say, but I know he'll deny it. He always does. Instead, I say, "Paul, there are different ways of showing anger. You just think if you don't shout or yell, you're not being angry."

By now we're at my apartment. I say good-bye and go upstairs alone. If we weren't in therapy, I'd end this now. Two days later, when it's time to go to judo, Paul calls. "I don't feel like going tonight," he says. So I go alone, and within a couple of weeks, Paul has stopped going altogether. I can tell he'd like me to quit, but I keep it up for a few more months before I drop out, too.

"See this," Paul says one afternoon in a department store, placing an Italian leather belt in my hands. It's an early spring day, buds are finally replacing snowbanks. "This is just like the belt the guy used to strangle my father. You should have seen his head. It was swollen and blue. It was grotesque. He was hardly recognizable."

Paul walks away, and I'm left to finger the soft black leather. I'm fixed to the floor, left to stare at his departing back, but all I can really see is his father's face—swollen, blue, grotesque. This is how he lets drop details of his father's murder. In the most ordinary moments.

"He was awake when he was strangled," Paul says casually one day as we walk on the path around the lake near my apartment. "The guy didn't drug him first." Just like that—no preface, no afterword—a simple statement of fact. He never wants to talk about it if I ask him to. He simply tosses out these bits for me to catch, one after another. In the most ordinary moments.

I never know what to say. But something, I have to say something. "God, Paul, it's all so awful."

"Yeah, well . . ." That's the usual response.

I can't tell how he feels. I can't read what this is doing to him. While such facts turn my stomach, what I mostly feel is sorrow. I can't bear to have anyone feel as alone as he looks.

Chapter 9

"I want to show you Africa," he says one warm May day as we sit in his living room. We're side by side on the leather couch, and spread across our laps is a large photography art book of Africa. He wants us to go there after the trial, which is scheduled for early August. He describes Victoria Falls, the largest waterfall on the planet, and the zebras, wildebeests, and lions he wants to photograph. He offers the trip as if it were a honeymoon.

Back and forth, we tick off places we want to see. In the months we've been a couple, we haven't traveled together, except for the occasional weekend trip. I yearn for time away from reminders of his father's murder, which I sometimes fear might rob us forever of a "normal life." I'm taken with his plans for us to travel to Africa, but I can't reconcile his two selves—the one who refuses to talk of his past, except in tidy scattered crumbs, and the one who wants to plan a detailed future together.

I tell him that I'd love to go to Africa with him, but that there's something about going so far away with him that makes me nervous.

"Are you afraid of me?" he asks, with an angry look.

"I wouldn't put it that way," I say. I tell him I don't feel as if

I know him. "Whenever I ask about your life before we met," I say, "you get angry."

Paul doesn't say a word.

"See. Like this," I say. He still doesn't speak, so I go on. "I've hinted often enough, and I've asked subtle questions about your past relationships. I've waited, and whenever you ask about mine, I tell you. But you never take the hint and talk about yourself."

I edge away from him and smooth my hands over the soft worn leather of the couch. Silence spreads between us. "If you don't want to answer, just say so," I finally say. "There's nothing wrong with asking."

How profoundly I hate sins of omission. I may have slipped from à la carte Catholic to ashamed-to-be Catholic (at least in the Church's current state), but my heart holds fast to doctrine that's universally sound. I find his secrecy and refusal to answer innocent questions frightening. I can see he doesn't understand the difference between privacy and secrecy.

Finally he speaks. "My past doesn't have anything to do with you, or," he adds, "*us*."

"Of course it does," I say, not even bothering to ask why he asks so many questions about my past, if that's his position about a person's history. Though it feels like picking at scabs, I always oblige, bleeding out pivotal scenes from my past.

"It doesn't," he insists in a thin distant voice.

I tell him I have to leave. Part of me wants to leave for good. But a larger part of me doesn't want to give up on him.

We fly to New York for Memorial Day weekend. I've booked tickets for *Dancing at Lughnasa* and *Death and the Maiden*. Being out of town doesn't ease the tension between us, however. As with most Irish plays, Friel's has plenty of laughs, but Dorfman's delivers nothing but realistic horror with the occasional attempt at humor.

For weeks, Paul continues to push for a late summer trip to Africa. Finally I tell him I won't go. "How about somewhere in Europe though," I suggest, striving for a compromise. He's clearly angry, but finally he plays along and says he's always wanted to go to Scotland. We settle on a three-week drive through the Highlands and islands. I buy travel books and start to map out an itinerary.

In the middle of summer I turn forty-two, and, after breakfast in bed, we celebrate with a garden tour, a visit to friends, a walk along a lake, and dinner at one of my favorite restaurants. In the garage is a new bike. His enthusiasm for pleasing me is overwhelming. When I thank him that night, he says, "You've given me so much this past year, it makes me happy to do something for you."

Then he asks me to show him some of the travel books I've bought. I've already put Post-it notes on several pages. For two hours we browse through the books, turning down page corners to mark places we want to see. On such days, I feel well loved.

Paul begins to withdraw again, and, instead of back and forth, it's all back—away from me. We spend most of our free hours together, but it's unclear why: he's unhappy and distracted almost all the time.

By now I've relaxed my no-sex rule, but it's not very satisfying for either of us, as far as I can tell. It's as if he's all out of joy. It's a question of trust for me, I tell him. He won't talk about it, but I'm beginning to suspect sex for him isn't nearly as exciting when intimacy's involved. In my cynical moments, I think: Great, after all this, we'll be emotionally connected, but the sex will be lousy. I try not to focus on that possibility.

* * *

The trial begins in early August, just eleven months after his father was murdered. Jen and Doug and Paul sit through every hour of the proceedings. Paul calls nearly every night, sounding oddly detached from it all. Jen calls a few times, too. She's worried about him. "It's tearing Doug and me up, but Paul doesn't show any emotion at all."

I don't tell her all the ways I see it affecting him. Instead, I say, "I know. I try to get him to talk, too, but he won't."

We keep waiting for it to hit him, she says. "I don't know what he'd do if he were alone," adding how glad she is he's with someone as strong as me.

The verdict comes eleven days after the trial starts. He's found guilty. The sentencing is set for November. When Paul comes home, he's as distracted as ever. I tell him I'm considering not going with him on our trip. "I just don't feel comfortable," I say.

"I know I haven't been myself," he says, "but we'll spend lots of time together before we leave for Scotland. You'll feel better then." But the week before we're scheduled to go, he flies to Florida to help with the Hurricane Andrew cleanup, and he comes back just in time to catch the plane to Glasgow.

I get on the plane anyway. By now our pattern is set: I talk about leaving. He talks me into staying. And I finish the job by convincing myself he's right.

Chapter 10

As our plane clears the shores of the East Coast, I leave my reservations behind. As soon as we land in Scotland, I feel at home; it seems one great-grandmother from this land is all it takes.

Most of the roads on our route are narrow, often bound by both loch and sea. Paul does most of the driving, and he finds the challenge of staying on the left side of the narrow, winding roads exhilarating. We have a list of possibilities but no set schedule. This type of travel—each day an open book—suits us both. The days pass with ease. Paul acts like the man I've come to love, tender and attentive. We tramp up and down heathery hills, and unpack picnic lunches of farmhouse cheese, smoked salmon, fresh-baked wheat bread, and Highland Spring water. At night, we stop at inns on the edge of the sea. We even talk of buying a cottage here.

A few days into the trip, we drive to a castle near the top of Skye. We give the castle itself only cursory regard; it's the gardens we've come to see. To call this botanical paradise a garden is to call a lion a cat. If Bergman had ever captured the purity of family angst in Scotland, he'd have chosen a site such as this.

It's rained all day, the kind of constant wet the Scots call mist, something more than fog and less than rain, but we don't mind. Like all seasoned travelers, we arrived prepared for rain by any name, with waterproof jackets and hiking boots. We find the waterfall and wander around the many gardens. For a while, we walk side by side, and sit together on a stone wall surrounding a fountain that's been here for hundreds of years. Then we take separate paths around the fountain, and, when I turn around to place Paul, I find that he's got his camera out. For the next several minutes, the lens is aimed at me, from different angles. I've never liked having my photograph taken—and he knows it—but it occurs to me now that it is petty not to cooperate. His contact sheets must be the equivalent of my journal entries, simply different ways to record life.

By the start of our second week in Scotland, we've reached the Hebridean islands of Lewis and Harris. Top of my list are the Callanish Standing Stones, an open-air cathedral at the edge of the sea. I walk down the row created by the highest stones and stand in the center place, most likely reserved for the high priest thousands of years ago. I breathe in the sea air, sweet and smoky with heather and peat.

Paul takes yet another photo of me. As he adjusts the lens, he says, "Don't look up until after I've taken the picture, but right over your head is a rainbow. You can see both ends from here." I stand perfectly still. "It's a sign," he says, as he lowers his camera and walks toward me. "If you'd been alive when they built this, women would have been in charge, and you'd have been the chief." He kisses me and turns me to face the rainbow. Sunlight washes over the ancient ground.

Though we've known each other for only just over a year, some moments I think I can be happy with Paul for the rest of my life. This is one of them.

* * *

Then the trip turns on us. I'm driving. Back on the mainland, we're on our way to an inn on the shores of a Highlands loch. It's been a long day in our small rented Volvo. The scenery outside is stark, with heartbreaking beauty, mountains and lochs, forests and sea. Inside it's stark, too, but with anger.

A minute ago, Paul reached over to stroke my face and tell me how much he loves me, how he can't imagine life without me. As he withdrew his hand, he told me how sad he is that he isn't sexually attracted to me anymore.

"If only you were more of an athlete," he says now, "I'd be more attracted to you."

This infuriates me. "You were attracted to me when we first met," I say, "and nothing's happened to me since then. I look exactly the same. The only thing that's changed is that we're getting to know each other. And you love me. At least that's what you say," I add. His words wound me, but mostly I'm angry.

He hates logic at such moments. He sees he needs a new tack. "If you'd only think of more fun things for us to do," he says in response. His eyes lock into a stare.

I remind him that I planned every single detail of our vacation. I did the research. I mapped the route. I drew up the list of quaint inns with spectacular views and out-of-the-ordinary sites. Once again, I'm glad I insisted on paying for half the trip. Never mind that everyone assumes that he, the multimillionaire, always picks up the bill for our travels. In my insistence to be equal partners in this area of our life, I often pay for more than my share.

I'm disgusted with myself for being so wounded when Paul acts this way. By now I'm a pro when it comes to absorbing hurtful words. At least I've picked someone who uses only words instead of words *and* fists. That's how I console myself. Hard as this is, it doesn't compare to my father's last rampage—thirteen years ago, but still fresh in my mind. What does feel the same is how I keep getting back up. I simply refuse to stay down after each new blow.

And, as my father always did the morning after, Paul, too, follows with expressions of love. On the heels of his if-only litany, which lasts the rest of the afternoon, Paul tells me how much he loves me. "More than anyone I've ever been with," he adds. Then, amazingly, he talks about how he'd like to have a family together.

My primary thought is escape. I should pull the car to the side of the road, grab my duffel, and hitchhike to the nearest train station. That's what I think. Instead, I take a deep breath and ask the question that I—that we—can no longer ignore.

"Do you think it's possible you were the victim of incest, Paul?" Our therapists have suggested this as a possible explanation for his behavior in intimate relationships. Paul dismisses the idea, but the longer I'm with him, the more I wonder if it's true. Just thinking about the possibility makes me feel sick to my stomach, and more sorry for Paul than ever. Some days, it makes being angry with him almost impossible.

Paul doesn't answer me. He simply stares. If eyes are meant to be the gateway to the soul, I've hit a blockade. When Paul is angry, his eyes bully me into silence. If I return the stare long enough, which I'm tempted to do now even though I'm driving, he looks away, even angrier than before.

"Please talk to me," I say in as even a voice as I can manage. I wait. "Paul," I begin again, "I think it's a possibility. Everything points that way." This time, I don't stop. "I won't consider having a child with you until you sort out why you're so angry and cruel to someone you say you love. I won't subject a child to that."

"I'll tell you what makes me angry," he says in a tight, controlled tone. "You want *me* to be the problem. You want to think that I'm disturbed and that's the reason I'm sometimes repelled by you. You don't want to think that *you* could be the problem."

"You know that's not true," I say, trying not to give in to my anger. "I'm plenty fucked up, but at least I admit it and work on it. I'm not denying what I come from."

Now it's my turn to fall into silence. I'm close to bursting. My body doesn't feel large enough to hold both anger and compassion; the car itself seems stretched beyond its limit. In our therapy sessions, I'm always the first—usually the only one—to take responsibility for my failings, our problems. He sits silently, coming forward only when a therapist aims the light of inquiry in his direction. By now, they're no longer attributing all of his unhappiness to the murder: they let him know he clearly had preexisting problems.

"There's something going on that I can't understand," I say, trying to keep my voice steady, even though I'm shaking with anger and fear, "and if it's not *that*, then what is it?" I glance at him. But he's still looking away from me.

"I've already checked it out," he says, "and it's not the case."

I don't buy the ease of his declaration. "How and *what* did you check out?" I ask.

"It's none of your business," he says.

"It is if you want a child with me," I return, and wait.

In a few minutes, he says in a voice so low I have to strain to hear, "I have a memory, but there's nothing behind it. Leave it alone."

I pull into a side road. "A memory of what?" I ask.

"Leave it," he says.

"A memory of what?" I repeat.

He aims his words outside the car as he tells me he remembers feeling something being forced into his mouth when he was a child. "It felt like I was choking, like I would gag. I was little, and I couldn't do anything to stop it." His voice is dull, but his anger is sharp and, it seems, pointed at me.

I feel sick to my stomach. "My God, Paul. That sounds awful. I'm so sorry."

"Don't be sorry. Nothing happened," he says, almost in a whisper.

Neither of us says anything else for a minute. I want to reach over for his hand, but am afraid to. Often when I touch him gently, especially on the back of his neck or on his face, he jerks away, as if something unpleasant is sure to follow.

If you want to preserve most of the membrane when you peel an orange, you run a slender spoon under the bright orange grain. A silver sheen runs just ahead of where the steel connects to the fruit. It's as if the fruit is being given notice it will soon be devoured. That's how it is now whenever I offer Paul a tender touch. It wasn't that way at first, so I don't always catch myself in time. Even during sex, it isn't tenderness he likes. It's as if he doesn't trust gentleness, or can't register pleasure unless it's close to pain.

So now, in the stopped car, I know this is no time for touching. I try to convert my love into words. "That sounds like what I'm talking about, Paul," I say after a minute.

"*Nothing* happened," he repeats, still staring straight ahead. "It's just a feeling, and nothing's behind it." His face is flushed with rage.

"Will you check it out again?" I ask him. "It doesn't sound like nothing to me."

"No," he says emphatically, as if I'm a dense student. He doesn't look at me while he speaks. "It *didn't* happen. *Nothing* happened."

All I know is that you don't hide "nothing" as ferociously as this. Once again, Paul's past feels impenetrable.

* * *

That afternoon, when we get to the inn I've chosen for the next few days, I walk down to the shore of the loch. I sit with my back against a rock as high as a chair and try to read. Unable to focus, I dangle the book between my legs and stare at the loch. Overhead, thick layers of gray cloud roll over one another, like a shroud moving in to blanket the deepest sleep of all. I hear nothing but my breath, and an insistent wind stirring up the lake.

I think of Paul, a wounded boy stuck in the past. Ever the optimist, I tell myself that he wants to become a man who learns how to recover from his injuries. On the shores of this Highlands lake, I will myself to be gentle and kind to the person I love.

As usual, within an hour, he seeks me out, sits down next to me, and wraps me in his arms. "Forgive me," he says. "Please." He burrows his face in my neck. "I love you." After a few minutes, we walk back to the inn, a sprawling wooden building full of warm fires, bright chintz armchairs, and people who look like they hide nothing but the most innocent of secrets. I wonder how we look to them.

That night, in bed, Paul reads to me. It's a ritual we rely on to smooth jagged hours. Before we left, I'd assumed that, freed from external reminders of his father's murder, we'd return to the relaxed, playful sex of our first weeks together. Instead, Paul rejects any overture I make. I can't help but feel his rejection as my punishment for insisting on therapy, not to mention for conversations like the one we had in the car today. He's all for talk, as long as I'm doing most of it—and as long as it doesn't touch on his past. And he wants physical closeness, but not sex. I lie with my back pressed tight against his chest, and I know he won't relax his grip until he dozes off.

Long after the pattern of his breathing tells me he's asleep, I drop off and dream: we're caught in a storm in the middle of the sea not far from our window. He's drowning. I'm the lifeboat.

Chapter 11

I can't always keep my panic at bay. After dinner a few nights later, in an inn miles from the nearest ferry or car rental or taxi, I lie beside him in bed, between crisp cotton sheets and a thick duvet. Even though our room is full of the cool, clear air of the Atlantic Ocean just outside our windows, I feel as if I might suffocate. I try for one good deep breath. I'm beginning to feel this way anywhere—in the middle of a street, a restaurant, or in bed. My body can't deny the stress of our constant storms.

Earlier in the day I asked Paul why he resists making love. He refused to talk about it. He hardly spoke during dinner. Silence can be comfortable. It can be friendly. Or it can be spiteful. His wordless rejection feels like a weapon against which I'm defenseless.

I will myself to breathe even though the air feels thin. If I close my eyes and Paul doesn't touch me, I can imagine myself alone—safe, ready for sleep. I picture a room: pristine, clean, housing no one's feelings but my own. Often, when he hears me struggling for breath, he doesn't say a word. He simply pulls me into a tight hold. This stranglehold never leads to sex. It's a goal, not a means.

If he does that tonight, I'll suffocate for sure, I think, finally recognizing this as a panic attack. I can hear him rolling over. As he reaches for me, I slip from the bed and throw on my clothes. He doesn't say a word. It's midnight when I walk out of our room into the village street that fronts the shore. I'm met only by the light of a lone streetlamp and a distant moon. Braced by the ocean air, I walk past whitewashed cottages with sleeping facades until my breaths are regular and measured, until I'm tired enough to sleep.

When I get back to the room, I can tell by his breathing he's still awake though he doesn't say a word. Our eyes meet in the faint light from the streetlamp outside our window. As soon as I get back into bed, he reaches for me. "Not tonight," I say. "I want to be left alone."

"I love you," he says. "I love you more than I've ever loved anyone."

I can't bring myself to say anything reassuring in return, even though it makes me feel like a rotten human being. His words don't feel like comfort. They feel like a vise. These days his high tides of affection leave me wrecked.

On the ferry to the Orkney Islands, I feel sick to my stomach and wonder if it's being on a choppy sea or being with Paul. By this time, we're swinging back and forth between empty attempts at cordial conversation and miles of hardly speaking. The next afternoon we drive to Scara Brae, the village unearthed, nearly intact, to reveal life as it was more than five thousand years earlier. A cold wind rolls across the water and a gathering fog smudges the clean lines of the village homes. Here people lived almost as we do today, only with less space, less luxury, less everything. They displayed delicate pottery in built-in dressers and cushioned sleeping platforms with grass. The hearths, cold for thousands of years, still hint at warm bodies roasting meat and fish.

I mentally pack up every image I can, every path, every hut, and every blade of metalwork. I want never to forget this singular place on the planet. I slow my step, unwilling to rush.

Paul moves quickly through the place, dropping his eyes at the site every now and then, and turns toward the car park.

"Are you cold?" I ask.

"No."

I watch his red jacket fade into the thickening pink mist. Did he even look at the sea, at the homes still intact, at the heartbreaking expanse of sky? He wanted to come to this place, but can't wait to get away. He stands beside the car like an impatient taxi driver waiting for a lingering passenger.

Even the day looks lonely: the sky is pressed so close to the earth I can barely see the heavy waves of the North Sea. But I can hear the salty water pound the beach and then drag itself away. If I fall into these waters, I think, I'll simply float down to the Irish Sea and wash up on the shores of another time. That's how lost I feel.

Chapter 12

Through the French doors off Paul's bedroom, I see nearly bare oak limbs in the backyard and wonder how long a person can keep hope and love ready, and feel only foolish for it. It's Saturday afternoon. I've been back from England for a few days; when Paul flew home, I went on to Yorkshire for more Brontë research. Now that I'm home, he wants to be with me nearly all the time, yet he argues with practically everything I say.

Paul's eyes are closed, but he's wearing his usual after-sex face. It's been a while, but I haven't forgotten that face—relaxed but distant. He thinks I haven't any taste for impersonal sex. He's right. I don't.

"I feel like a replaceable part with you," I manage to say.

He doesn't move, but his mouth tightens. I suppose he's getting ready to defend himself. I wait to see what shape it will take, silence or carefully constructed rage.

With eyes still closed and a voice barely audible, he says, "What is it you want from me?"

I sit up and position myself on the edge of the bed. I reach for the black comforter that has slipped to the floor and wrap it around myself.

"I'd like it if you opened your eyes when I talk to you." It feels crazy, having to ask this of a grown man. "I feel invisible when we have these conversations and you won't even open your eyes."

I wait. It's like talking to a pouting three-year-old. I roll my shoulders and sit up straight as I practice patience. After a few minutes I say, "Please say something."

"What do you want me to say?" His eyes are still closed.

"I don't know. I only know what I want to say. But I want you to say *something.* And to open your eyes."

He speaks. "You want too much. Love isn't supposed to be this hard."

"Says who?" I ask. "Where did you learn that love's supposed to be easy? It's only easy if one person is in charge and the other just follows the rules."

He works his jaw and opens his eyes. I catch a flicker of thought: Was there ever a time his eyes held a smile? Or did it always stop at his lips? These are Modigliani eyes, blank. Here it comes, I think.

"Maybe we should just admit that if it's this hard, we're not meant to be together," he says.

It's amazing how you can love a man and find his thought process completely alien. Love is the hardest thing of all, I want to shout. Instead, I say, "Maybe. But if I leave, it won't be because I believe hard means it's not meant to be. If I leave, it'll be because I'm with someone who fades at the moment he could make himself just a little bit vulnerable."

Now he's looking at the ceiling. He says, "What do you mean leave? *You* won't leave. If anyone leaves, it'll be *me.*"

I let this sink in. I change my mind. Now I just want him to shut up. But he keeps on going. He tells me it's unreasonable to talk about a lifetime of getting closer and getting to know someone. You get to know someone, and you settle into a comfortable pattern, he says, adding, "I don't want trouble."

I try to stay calm. "But love is not easy and comfortable all the time," I say quietly. "Sometimes it's about the most uncomfortable thing in life."

"I don't like anything that's not easy," he says. He repeats that he thinks we should admit we're not meant to be together.

"Life can be hard. *Love* can be hard," I say. "Happiness isn't something that just happens to you. It's something you *do*. It's something you *make*. It's not a prize you get for picking right." Well, picking's part of it, I think, but I don't say it aloud.

He finally looks at me, with eyes set cool. "I don't want to fight," he says.

"Then why did you pick someone who's willing to fight when it matters? The first thing you loved about me is how hard I'm willing to push against what seems wrong to me. You like being with someone who's willing to take on the whole world. Well, the whole world includes us. I'm fighting for *us*!"

He eyes me warily.

"You like being next to someone who does all the fighting," I say. "It saves you doing any of it yourself. But when it's me—or us—I'm fighting for and it has anything to do with you, you disappear."

Now he rouses himself. "You think you're so smart. You think you know me so well."

"Maybe I don't know what drives you, but it doesn't mean I'm wrong about this."

Now I'm the one who looks away. After years in the glare of my father's rage, my feelings have an antiscald sensor. When I can see someone's rage coming white-hot toward me, I close the valve to my heart. It acts as a two-way device: it blocks what's coming toward me, but it also caps what I feel. Sometimes it serves me well. Sometimes it keeps me too controlled. Part of the control is my determination never to scream hateful words that cannot be taken back. Growing up the way I did, when rage was not followed by sorrow, makes you afraid to repeat the pattern.

So instead of saying what I'm tempted to, I reach for my robe and start walking toward the bathroom. His silence draws me back. I'm disappointed in myself even as I turn back and sit on the edge of the bed again.

"This isn't how we want it to be, is it?" I say, leaning toward him.

He looks at me and strokes my cheek. "You worry too much," he says. Turning his gaze away, he adds, "I love you, but this is too hard." I can barely make out the words.

The worst thing about his "I love you" is that it's beginning to feel like a short rope thrown from a departing ship. I reach out and rub his thick, wavy hair. I'm reaching my tolerance for feeling ignored and invisible. I try one more time. "Please look at me," I say.

Even then he doesn't. Instead, he turns and works his head into the opening of my robe. I hold his head above my breasts. I rest my cheek on the top of his head. At this moment, holding him like this is less lonely than looking into his eyes.

When I leave his house a few minutes later, I mean to leave for good. I go home, cry for a while, and dress for a friend's wedding. I reach for my cobalt-blue dress and give myself a good swipe of lipstick. I refuse to look as upset as I feel. Standing in the midst of friends in a downtown sculpture park, feeling both relieved and angry, I watch a hopeful young couple enter that mysterious space created when two people pledge fidelity to each other.

* * *

By noon the next day, I have several messages from Paul, each one sounding more urgent than the one before. Part of me is curious. Part of me is flattered, though I'm not proud to admit it. When I finally pick up the phone in the afternoon, he asks me to meet him for a walk.

"I don't want to lose you," he says as soon as we start around the lake path. "Come live with me."

When I start to protest, he puts his hand over my mouth. He insists it will be easier if we live together. He tells me he loves me, more than ever, more than anyone. He pulls me toward him. I can't see his face any longer. "Please don't leave me," he says. "You're my only chance."

"Of what?" I ask.

"Of not turning into my father," he says.

These words—and his pleading look—go straight to my heart. Still, I can't help but ask what will change.

Can you imagine, he goes on, what it's like to have your father murdered and to discover along with everyone else that he had a secret life? He doesn't wait for me to answer. "It won't always be like this," he says, assuring me that with the trial over, life will get back to normal. "It'll be better for me—for us," he adds. "I promise. You've got to believe me."

By now, I've pulled away so I can look at him. When his face looks blank, I feel nothing for him. But when he looks like this, desperate, pleading, like a child, I can't resist.

Besides, what he says is exactly what our therapists are saying. It makes sense. My doubts aren't completely silenced though, even when he tells me how sorry he is for all the times he's been distant and cool and cruel. "My father's death," he says, "is affecting me in ways I don't like. I'm not always the man I want to be." He tells me he knows he needs private counseling and promises to get help. (While he has been showing up at our joint sessions, he has slacked off on his own.)

By dark, I allow him to convince me that my answer should be yes. Aside from counseling, I have only one condition: I insist on assuming some of the expenses of our living together. I refuse to live off his money.

If this were a scene in a movie, I'd be groaning. Why do I say yes? My past cancels out my logic: I talk myself out of my doubts. This feels so familiar: whenever my mother drew the line and sent my father away (never far, to the cabin or to a hotel), he always talked his way back within a few days.

Like my mother and her mother before her, I stay with a difficult man because I can't bear to abandon a wounded child. And because, in my heart of hearts, I want to think he'll be like my father: a stubborn but not hopeless case.

Chapter 13

As if to blot out the past year, Paul turns gentle and kind, engaged and present. He seems content, which is how I feel, too. Our nights are better than any we've had so far.

To lock the door on my old life, I give up the lease to my apartment. I can't allow myself a way out if I'm going to give this relationship a fair shot. As soon as I move in, Paul starts talking about getting married. I don't encourage the conversation. Sharing a house feels foreign enough. I've never lived with a lover before.

And the truth is, I've never planned on getting married. When I was young and my friends fantasized about weddings, I was dreaming about London and Paris and Nairobi. When I was six, my maternal grandfather went on a grand tour of Europe. In addition to the postcards and presents—a proper tea set from England, for one—I kept his memories. I heard his stories so often, I sometimes forgot I wasn't there myself. It's as if I sat next to him on the hotel veranda in Naples drinking espresso and eating gelato while he waited for a shipment of his beloved Cuban cigars. I can see him on planes and trains, walking the broad avenues of the same Europe he fought to save forty years earlier.

Such stories direct the way I turn at every juncture of my

life—always toward the world. The closest I've come to moving in with anyone was years earlier when I committed to a schedule with Jim: three nights a week at my house, three nights at his, and, at my insistence, one night apart. He was the only man I ever seriously considered marrying.

Perhaps this relationship with Paul is the way I'm meant to learn the difference between adoration and love. I feel confused. I know what need feels like, but I'm not so sure about love. I'm not sure I'll recognize it, but I'm determined to try. Even so, when he talks of marriage, I discourage it. "Let's just try living together," I say. "I may not be any good at it."

Besides, Paul has been divorced twice, I find out shortly after I move in. He finally tells me about his first marriage, which took place when he was in his early twenties. His marriage to the second ex-wife—she of the vanity bathroom—ended several years ago; I'd heard all about it from Jen. Even though I don't know many details, I know enough to see he was at least partly at fault. And yet, our life together seems to be heading into a place where I can imagine being happy for a long time, maybe for life, married or not.

We're full of plans. Paul's open to changing the house in almost any way that will make it feel more like home to me. I shop for new bedding and replace the black comforter with a cotton duvet the blue of a Highlands loch. I buy a light maple table so I can work upstairs while the dining room is transformed into my office and a library. I make meager progress in the living room. And I never quit wishing for and requesting additional interior doors. This house has hardly any doors—plenty of openings into rooms and door frames prepared for doors, but no doors. Paul seems to prefer extreme privacy outside the house, but little once he's inside.

Most important to me is that Paul plans his weeks around

twelve-step meetings. He doesn't talk much about them, but I gather they're for children of sex addicts. And he's taking the course work he needs to finish his undergraduate degree. He's bringing meaning and structure to his life, which I read as a good sign.

We develop an absurd hobby: cooking. Absurd, because it involves little actual cooking on our part. Instead, we *talk* about cooking; we *think* about cooking; and, our favorite bit of all, we *shop*, scouring kitchen stores for gourmet utensils. Props, really, since mostly we eat out or order in. Once we get into a store, we drift apart. I am riveted by whisks: balloon whisks, French whisks, ball-tipped whisks, flat whisks. And anything having to do with the preparation of tea: pots, caddies, filters, cozies. And pitchers. I *love* pitchers. They're good for everything, from iced tea to tulips.

Meanwhile, Paul usually heads for the blenders and rice cookers and fish platters. He *loves* fish platters. We find a perfect one for the salmon Drambuie recipe I brought back from a restaurant on the Isle of Skye.

We spend months looking at food processors but can never agree on which one to get. Which doesn't matter much, of course, because we'd hardly ever use it anyway. And while we're eating in restaurants and cafés, we discuss the chef's cooking and consider whether it's something we could replicate, you know, if we actually ever cooked ourselves. Just to keep my dignity, I cook once or twice a month.

We take this shared hobby very seriously, yet we often end up laughing hysterically at the silliness of talking so much and doing so little: buying utensils because we like their looks, even if we seldom, if ever, actually dirty them.

* * *

It's finally time for the sentencing. We fly to Albuquerque, and Jen and Doug pick us up at the airport; we're staying with them at the town house Roger used as a guesthouse. He left it to them.

The day of the sentencing is sunny and dry, perfect southwestern weather. We should be sitting at an outdoor café ordering lunch instead of inside a courtroom waiting for a convicted murderer. When we get there, Doug leads us in, then stands aside and says to Paul, "You first." Paul moves ahead of us and sits down. I follow. Jen sits on my left and Doug sits down at the end of the row. As I look up at the prisoner entering the courtroom, I understand Doug's intent: he wants Paul as far from the man as possible.

I reach for Paul's hand. It's cool and damp, reminding me how much this is affecting him even if his face doesn't show it. I force myself to hold on through the entire court session. I want to ground Paul to something other than what is happening in this room.

When the judge asks whether she should consider any extenuating circumstances before she delivers the sentence, a woman in a tailored dress and sandals takes the stand. It's the man's sister. As she answers a few preliminary questions, Jen leans toward me and whispers, "She looks like a nice woman. This must be awful for her, too."

I nod my head in agreement as the woman makes an articulate plea, describing the difficult childhood she and her brother shared. The judge thanks her, and, when she walks back to her chair, she turns our way. She mouths the words, I'm sorry. I'm *so* sorry.

When it's time for the judge to read the sentence, she singles out the factor she can't get past: no remorse. Not one sign of remorse in all the days of the trial, and not one sign of remorse today. That's what it comes down to. The judge is not inclined to put more emphasis on a childhood, however terrible, than on a heinous crime. Besides, she points out, the same childhood produced his sister, who by all accounts is a loving mother and a solid member of her community.

The sentence is the maximum allowed—death. It's what Roger's family hoped for. I don't believe in capital punishment, but I know this is not the time to share that sentiment. Later that day, Paul is giving a party at his father's country club for all the people involved in the investigation and the trial. Jen and Doug think it's a wonderful idea. But it seems strange to me; I'm not entirely sure what we're meant to celebrate. I feel out of sync with the feelings around me.

That afternoon, in the little time we have back at Jen and Doug's town house between court and the country club, we sit on the patio. Paul asks me to help him think through what to say that night to thank everyone. Jen gets pen and paper, and I help Paul write out a short speech.

When we're finished, I decide to take a shower. I need to do something to rinse off the day's dirt. Doug and Jen stay on the patio, and Paul follows me down the hall. "Want company?" he asks. We're quick. And quiet. He starts to say something. I hold my hand over his mouth. This time I'm the one who has nothing, absolutely nothing, to say.

Late one night not long after we get home, my phone rings. We added another phone line so I could transfer the number I had at my apartment. I'm in the bathroom washing my face. Paul is in bed. He gets up to answer it, perhaps forgetting it's my line.

I come out of the bathroom. "Who is it?" I say. "It's late."

"No one was there," he says, back in bed now. Just then the phone rings again.

"I'll get it," I say, recognizing my ring again.

"Hello."

Silence.

"Hello," I say again, and it occurs to me that this is the time of night Jim usually calls, and I haven't heard from him for a while.

Nothing, but I know someone's there.

"Must be a wrong number," I say when I put the receiver down, though I'm almost certain it was Jim. This would have been the first time he's called since I moved in with Paul. He knows about Paul, but not that we're living together. I say nothing more, and Paul doesn't ask, though the way he watches me walk toward the bed suggests he'd like to.

I'm not sure how I feel about this. I want to talk to Jim, but my habit is never to call him, which is how I protect myself from giving in to one of his proposals when I feel lonely—or just plain miss what we had together.

Chapter 14

A few nights later, as we stand in the kitchen discussing our options for dinner, Paul asks me the time. He never wears a watch, so I'm the timekeeper. When I tell him it's just after five, he says, "Let's call your sister and the children. I want to say hello." It's about the time she usually gets home from picking up the children from day care.

Five o'clock to six o'clock, I tell him, is the hour from hell when you've got little kids. I suggest we call later, but he insists, so I humor him, go into the dining room, and dial my sister's number. When Liz picks up, I can tell something's wrong.

"Everything's okay now," she assures me when I ask, "but you're not going to believe what just happened." That afternoon, she proceeds to tell me, she stopped at a convenience store for milk after picking up the children at day care. "I've never left them in the car before," she says, sounding distressed, "but they'd fallen asleep, and I hated to wake them." She parked directly in front of the store's glass doors and locked the van, knowing she'd be able to see it from the refrigerator case.

When she got to the cash register, the man stepping in line behind her said, "Is that your car?" and pointed to a man who was trying to get into her van. She dropped the milk and ran outside. The man jumped back into his car, which was parked alongside hers, and drove away. The children slept through it all.

She shouted for someone to call the police. When they arrived, she explained what had happened and gave them the license plate number of the stranger's car. Despite her requests to have them fingerprint her van, the police told her it wasn't necessary. Her insistence, however, resulted in their putting out an APB. She took the children home and waited to hear from the police.

"Can I get on the other phone?" Paul asks. He can tell something's wrong. I fill him in, and he picks up the kitchen extension. "Did you catch the license plate?" he asks Liz. When she says yes, he tells her he has a friend who can get into computer files and check license plates faster than the police. "He's one of the best computer hackers in the country," Paul says. "Will you let me help?"

Liz says she's afraid the police aren't taking this seriously, so she's grateful for his offer. Paul gets off the phone, and I can see from the light on my phone that he's switched to the second line.

Just a couple of minutes later, Paul comes back into the room and takes the phone from me. "I've already got an answer about whose car it is. You're not going to believe this, but it's someone I know." He tells her it's a man by the name of Gil Greenwood, a man who used to work for him. "Until I fired him," he says, and explains he's always been afraid someone would try something like this to get money out of him. He speculates that Greenwood knew how close he's become to Liz and Andy's children, and thought Paul would pay a lot of money to get them back from a kidnapper. I'm horrified by the idea.

My sister assures Paul that it's not his fault. I echo her thoughts. After we hang up, Liz calls the police and passes on the information Paul gave her. When she calls back later, she tells us Greenwood's wife told the police he was out of town on business. She didn't know where, but she expected him back home that evening.

The police find Greenwood in a small town fifty miles away from my sister's home. His story is just that, a story—more accurately, a few stories, as his original alibi doesn't remain consistent during the hours he's questioned. First, he claims he wasn't ever in the small town where my sister lives. When the police tell him he was seen and his license plate recorded, his story changes: the children were crying and he wanted to make sure they were all right, he says. But since they were still asleep when Liz got back to her van, this is clearly another lie. It's his story against my sister's, and since the children weren't harmed, the police let Greenwood go.

When I talk to my sister later that night, she tells me that prior to this incident, a man had called both her school and the United Way, where she serves on the board. Each time, he asked about her schedule. My sister assumed the calls were from the father of a student; now she thinks they might have been from Greenwood.

A few days later a friend of my sister's who is on the local force calls to tell her that he believes her story, but that the police chief and the rest of the department aren't taking it seriously. The county attorney refuses to prosecute.

Chapter 15

We spend the end-of-year holiday with my family. The night after Christmas, my father asks me to take a walk with him. As we head out into the dark, cold air, I think of all the walks I took with him as a child. What I remember most is reaching my arm up to the large hand waiting for mine. And I remember—as I stretched out each step to keep up—how our feet crunched the snow cast blue by the town lights. I suspect he closed up his own step so that I could match his stride.

These walks were a rare chance as a child in a large family to have my father's undivided company. He taught me on those walks about the northern lights. We'd stop in the middle of the sidewalk, and he'd point to the undulating streaks and tell me how lucky we were to live in such a place, where the sky puts on such a show. I remember how safe I felt, with my father's weathered leather glove cupping my thick wool mitten as we watched the shifting strips of light color the distant sky.

And then it all went wrong. For much of my adult life, I avoided being alone with him, a defense against his alcoholic rages. Now, years of sobriety and recovery from a massive stroke have conspired to mellow him, and we've been working our way back toward each other. We're approaching peace, but I haven't forgotten, nor entirely forgiven, his wounding rages. Yet I can see every time I'm with him now that his one aim is to return to the kind father of my childhood.

Since his stroke, he's walked with a cane; this time I'm the one to slow my stride. I hope I do it as gracefully as he once did. We talk of small things for most of the walk, and as we approach the house, he says, "I have only one piece of advice to give you. If you marry Paul, be sure you'd marry him if he didn't have a penny."

My father's a proud man. I'm the daughter who's made a habit of not seeking his advice. It would have cost him to offer it unasked.

His remark offends me. "I love him in spite of his money." My next words hint at a question. "You don't like him."

He expresses himself carefully: "I can't figure him out." I know what my father sees is the money and the enigmatic manner. "I've never known anyone like him. He says things I can't make sense of." This last remark is not a compliment. When it comes to people, my father is subtle about his likes and dislikes; the clues are mostly in what he doesn't say. It's his easy manner around those he likes and respects that lets you see who his heart favors. He's not easy around Paul.

I don't totally disagree with his assessment, but I'm too proud and stubborn to tell him that. It isn't that I need my father's approval. It's more important than that. I trust his judgment, his read on people. He's discriminating, but kind. He's unlike some in my family who aren't at all discriminating in the company they keep, or others who like people for all the wrong reasons. Not my father. He has other faults, especially when drunk. But sober, he knows how to read people.

This conversation unsettles me, which, of course, is his intent. But I convince myself he just hasn't given Paul a fair chance.

* * *

"I still can't believe Jim's dead," my friend Alex says one afternoon a few weeks later. We're having a late lunch. Alex and Jim and I have known each other since we were in our twenties and worked at the same law firm.

I nearly drop the glass that's halfway to my mouth. I put it down too hard, and ice water splashes onto my hand. I watch it drip onto the speckled Formica tabletop before I look back at Alex. "Dead?"

"Oh my God," he says. "You didn't know. I thought you knew." And then he tells me Jim dropped dead of a heart attack in November. He was in Paris. It was early morning, and he'd gone out in search of something for a headache.

Right after the phone call I'm sure was from him, I think.

We're sitting in a booth in the back, and the diner's nearly empty now. I keep my hands in my lap, to hide the fact that I'm clenching them. I've never talked to Alex about Jim's regular proposals. Now I suspect Jim hadn't told him either. Most likely, he thinks Jim and I are just old friends who parted on friendly terms when we were no longer a couple. Alex starts reminiscing about the days when we were all working together, when we were young and felt as if we had forever. I laugh in all the right places and try to conceal my sense of disorientation. All I can see is Jim falling—all six feet six of him—and cracking his head on a cobblestone street.

It's a typical Midwest January day: the bright sun belies the temperature, which peaks at several degrees below zero. In such weather, one becomes practiced at parting inside a building. As we hug, he says, "I'm sorry you didn't know before."

"It's all right," I say. "I'm glad you're the one who told me."

As I drive away all I can think is this: Jim is dead. So is our pledge to laugh together in the face of old-age loneliness.

Chapter 16

In March, Liz's children come for a long weekend. We pack in everything we can—movies, museums, bookstores, toy stores—and when they leave, we're exhausted. Once again, I'm struck with how good Paul is with kids: all else aside, he knows how to have fun.

The weekend is a welcome break; lately Paul has seemed discontented. I'm beginning to see that he knows much more about pleasure than he does about joy. While this new realization unsettles me, it doesn't prepare me for what comes next.

Several nights after the children's visit, Paul suddenly starts talking about all the things he hasn't been willing to share over the past several months. We've just come home from the video store—without a video. We'd had an argument. He wanted to rent *Basic Instinct*, which we had seen in the theater last year. One time was enough for me, I said, and made several other suggestions. *Basic Instinct* or nothing, is what he said just before walking out of the store.

It's a Friday night, and I've had a long week. Tired, I mostly just want to be left alone. When we get home, I crawl into bed with a book. A few minutes later Paul gets in bed. I can tell he's still angry.

"You think there's something wrong with me because I like that film," he says.

No, I tell him. "I think there's something wrong with you that you get this angry when I won't watch a film that disturbs me, and," I add, "when I want us to talk openly about our sexual relationship."

Without preamble, he says, "My last relationship ended when I flew to Thailand on the way to a photo shoot and had sex with a prostitute."

Fingers of fear crawl up my spine. "How old was she?" I ask. I assume it was a she. I don't want to interrupt him now that he's finally talking openly about his past.

He says he didn't do anything illegal.

"How young?" I ask.

His face sets hard, and he works his jaw. "Not a child," he says after a pause, "and that's all I'm going to say about it. It's too painful to remember."

This whole time with Paul, I've been looking at a picture made up of dots. With this one revelation, lines appear, and the dots connect. "Your meetings are for sexual addiction, not just because you're the son of a sex addict, aren't they, Paul?"

"I thought you'd figured that out," he says.

Then comes another bombshell. Before his recovery from sexual addiction, he says, he had lots of friends who were actively engaged in criminal activity. In fact, he and his father often hired people with criminal records at the family companies. If he ever quits his recovery program for sexual addiction, he continues, he knows he'll start living a life like his father's, an underground life of cruelty and sexual perversion.

I expected to be relieved when he finally broke his silence, but I find it only disturbing. And to tell the truth, alcoholics never looked better to me. Slurred speech, pouches under the eyes, shaky hands. I know how to tell if someone's sober. Or not. When it comes to sexual addiction, I'm a neophyte. And I don't like not knowing.

I come back to the moment. "But you *have* quit recovery," I say, for I know he's quit going to his meetings. I don't know when, I just know his schedule is free when it didn't used to be. By now, I'd suspected his meetings might be for his own sexual addiction as well.

He responds with the classic line of an addict who's flirted with recovery and wants to deflect discovery: "Don't take my inventory," he says. "It's not your business."

"It *is* my business if you go back to your former life," I say. I can feel the love in my heart shrink, to make room for fear. I feel battered by all this information, by his anger.

"You're going to regret it if you push me," he says. He aims cold eyes at me. They remind me of the night I met his father.

Long ago, I decided I'd never devote my life to an active addict. It's not addiction I fear; it's lack of sobriety. I believe in people's capacity to heal. I witnessed more than my share of binges and abuse. And then I witnessed a fair share of recovery.

The only way I know to cope is to slip into research mode. Over the next few weeks, I read up on sexual addiction. I talk to my own therapist. (Paul has quit going altogether.) It doesn't take long to discover that the rate of recidivism in sex addicts is higher than most chemical addicts. The truth is I'm not confident I'll recognize the signs when *my* sex addict slips, but I convince myself that I'm prepared to face any addiction. And I simply hope I'll know when he's "sober."

It doesn't take me long to figure out that Paul and I share the house with his parents' ghosts. Now that his father's been dead more than a year and a half, I feel free to tell him how unnerving I find the portrait in our bedroom.

"There's something unnatural about it," I say one day.

He laughs. "If this bothers you, you'd hate the image he had of me in his house."

I can't help my curiosity. "Tell me," I say.

And he does, describing a bronze sculpture of himself—naked, close to life size—commissioned by his father.

"There's something unnatural about all this," I say again.

"It's by a renowned artist."

"I'm not talking about the art. I'm talking about life."

He refuses to take the painting down, but I don't stop there. I also ask him to consider moving so we won't be living in the shadow of his parents' lives. The only place he's willing to consider is a condominium just outside downtown. I recognize the name of the development. It's where the woman he was with before me lives. He'd pointed it out one day. I tell him it would be cruel. For all of us, I say.

Well then, he says, there's nothing to talk about.

How about looking for a house near the lake just outside downtown, I ask.

He tells me he hates that neighborhood.

Another night in bed, he tells me he wants to buy me a baby grand piano for Christmas. He knows I love to play the one my parents have when I visit them.

"I don't want you to buy me a piano," I tell him.

"Would you rather have a car?" he asks.

"What?" I ask, confused.

"If you don't want a piano, would you rather if I bought you a car?"

"No, I don't want you to buy me a car. I don't *need* a new car."

A nicer car than you have, he says.

There's nothing wrong with my Honda Accord, I respond.

This goes on for several minutes. "Look," I say to close the conversation, "I don't need you to buy me things I can't afford myself."

But he isn't finished. "Let me know if you change your mind." He hugs me tight and holds me for a minute. "I love you, you know," he says. "I want to give you everything you ever wanted."

Before I pull away, I say, "I don't want a life like that."

When I look at him, he simply smiles, as if I'm a simpleton.

Another day, just as we're falling asleep, he says, "Why don't you quit working and I'll pay for you to go back to school."

"I've got my master's, and I don't want a doctorate," I respond.

When he persists, I ask him why it matters so much to him. We can be together more, he answers, if I don't have to work. Such togetherness doesn't appeal to me. I can't imagine life without work.

By the next week, it must occur to him to start smaller. "I'd like to get you a cell phone," he says.

When I insistently decline, he turns angry. I ask him why he wants it so badly.

"So I can always reach you," he says.

Paul's interest in sex wanes again. Whenever we're in bed together, he wants to hold me as if he's hanging on for dear life, but he acts angry if I try to initiate sex. One night in bed, he reverts to his old mantra. "I wish I was still sexually attracted to you," he says, apropos of nothing. "I love you," he continues, "more than I've ever loved anyone before. I'm just not attracted to you."

I feel his breath against my neck. I myself can hardly breathe. I feel as if I'm being suffocated. I say nothing. I'm tired of being logical about this.

"I hope someday that will change," he says into my silence, "but I don't know. I don't know if it ever will."

I pry myself loose from his grip. I get out of bed and sit on the floor, my back against the wood platform. The more I learn about sexual addiction, the more predictable this all seems. I'm not meant to take it personally. Everyone says so, but it *feels* personal. I don't know what to do. We've been here before. I've lost track of the number of times. I take a deep breath. "Do you find it strange, Paul," I say carefully, trying to put aside my hurt and confusion, "that the more you say you love me, the less attracted you are to me sexually?"

"I can't help how I feel."

"Do you think this has something to do with your sexual addiction?" I ask. "Maybe you don't like to have sex with women you feel close to." Or women who aren't needy, I think, deciding to keep that thought to myself.

I ask Paul to go back to therapy with me. So we go again, and he repeats his current refrain: He's never been as connected to another human being as he is to me. He loves me. He just doesn't want to have sex with me.

Exhausted as I am by this seesaw, I want to give him more time. And I don't want to have been wrong for loving him.

After all, in the midst of all this, he's doing what I've been waiting for: he begins to turn his dreams into a life that matters. He's close to getting his degree and is checking out medical schools. I help him whenever he asks: filling out forms, helping him think through and answer essay questions.

Anger at his parents seeps into all the crevices of this process, as if everything that's hard for him ends there. When I push him to talk about it, he tells me a story that makes me wonder what others he's holding back.

One summer evening, he says, before he was old enough to go to school, his father bought a pink-and-white dress and paraded him around in it, before the neighbors. It was a hot, muggy night, he recalls, and the dress was scratchy. Paul remembers the sound of his father laughing.

I wonder what his mother was doing. Was she joining in the laughter or simply sitting by silently? Perhaps she was more afraid of displeasing her husband than sacrificing her son. Surely she should have intervened.

Chapter 17

I never give up hope that being away from home will help us escape the pall of the past. So I encourage frequent trips. Paul thinks of himself as a world traveler, a connoisseur of adventure. He dresses the part—khakis and boots, tousled hair and tanned cheeks. He's still pushing for a six-week safari in Africa, but when I close my eyes and imagine myself on a continent where I don't know a soul except him, my neck and shoulders tighten. My answer is no. I don't tell him why. I just tell him no.

Instead, we go to Strasbourg to stay with a friend of mine. My friend, the consummate French businesswoman, meets us at the airport in a crisp white shirt and tan pencil skirt. We settle into her home with a shower and a nap, and then sit down to her traditional welcoming champagne. She's set a table in her garden, and we eat grilled steak and salad, finishing with the chocolate mousse she's famous for in her circle of family and friends.

As the days pass, I watch Paul turn into an ugly American. He never once says thank you, though my friend has gone to much trouble planning outings for us on her days off and has prepared lists of possibilities to occupy us while she's at work. I have to urge him to pick up the tab for dinner the night we go out. He could easily buy the restaurant. I have to remind him to pay the compliments any child would be mature enough to make, to show gratitude any decent person would naturally feel. He acts like he's at a hotel and he's owed good service.

At least once a day, Emilie and I exchange raised eyebrows, but she's far too tactful to tell me what she thinks of Paul and I'm far too ashamed to ask. Besides, she can see that I'm embarrassed.

Paul looks bored most of the time, though our days are packed. It occurs to me that he's a complete fraud. I'm left to wonder what kind of person says he loves art and then rushes from room to room, with hardly a second glance, as if there's nothing there he couldn't create—or commission, like the large abstract painting in our bedroom.

"Take your time," he says, but his impatience is clear. Art. Scenery. People. It's all the same. Never content to savor the experience in front of him, greedy for whatever might come next, he's insatiable. But no amount of money can make a moment hold the future as well as the present. Not for all the wanting in the world.

I see now that when we travel, I experience our trips alone. The difference is I'm happier when I'm actually on my own.

Paul and I drive into the French countryside, traveling south of Strasbourg along the Route du Vin. After two nights in modest village inns, we stay in a pricey hotel in Basel, just across the Swiss border. That night at dinner in the hotel restaurant, Paul becomes rude when the waiter doesn't answer all his questions in English. The waiter regretfully explains that he isn't fluent in the language. "When you pay this much for food," Paul says in an angry voice, "everyone in the place should speak English!"

As he speaks, I stroke the thick linen tablecloth and finger the silverware, heavy and prim against the white cloth. When he is finished, I smile apologetically at the waiter and say to Paul, "We're in Switzerland. They speak French and German here. And when Europeans are in America, they don't expect people to speak their language, no matter how much they pay for things."

Paul starts to argue. I ignore him and cast another chagrined look at the waiter, a man of about sixty, with thinning hair and an erect posture, and say, in French, "His behavior is inexcusable. Please accept my apology." I repeat the word the two languages share, "inexcusable," not caring that Paul will understand. "I should leave him," I say.

It took another language for me to try that out.

The next night, we have to choose between staying at an inn in the middle of a forest or in the middle of a village. I prefer the village, with streets full of visitors. We sit in the second-floor dining room of a traditional Alsatian restaurant. Paul is acting cool, looking at everyone but me. Yet when I ask him if he'd like to talk, he says, "Nothing to talk about."

I look around at the other diners sitting under the open beams and wonder if I'm the only one in the room wishing to be with someone else. I try to salvage the moment by imagining I am alone. It's less painful than trying to penetrate his stony anger. As I savor the creamy sweet onion tarte and the crisp white wine I've ordered, I recall my favorite part of the day: standing above the cloud line at Mont Sainte Odile, just one more pilgrim in search of a sacred moment. Paul, already in the car, was nowhere to be seen.

From this moment on, I try to avoid being alone with Paul. I can't sort out why—it doesn't seem logical to be afraid of him, so I feel foolish for it—but I trust my body, which tells me it will always be wise to pick the village over the mountain.

* * *

Flying home, we spend a night together in London. I'll be staying in London for ten days. I'm relieved I'd planned this research at the British Library. I need time on my own, away from him, to think. Early the morning of his flight, Paul wakes me as he leaves. He sits on the edge of the bed. "Will you be staying at this hotel?" he asks.

"I don't know," I respond.

"Why don't you stay here," he says. "I want to know where you are."

"I haven't decided yet," I say.

"Stay here," he says again, this time more insistently. He strokes my arm.

"I might stay closer to the library. I'll decide later." I'm still half asleep, but awake enough to see his eyes cloud in annoyance.

"Well, call me as soon as you know. I want to know where you are."

Often he doesn't even ask where I'm staying when I travel alone. This realization pulls me fully out of my pre-coffee fog. As I watch him walk out the door, I decide to stay somewhere else.

A couple of hours later, as I head off in search of a new hotel, I realize I'm being followed. In all my years of traveling alone, this has never happened. Several times during the next few hours, I slip into shops in an attempt to shake the man who's trailing me. Finally, I decide I've lost him. I retrieve my luggage, then check into a small hotel that's an easy walk to the British Library. It's too late to go to the library, so I decide to walk to Silver Moon, my favorite bookstore on Charing Cross Road. Coming toward me as I cross busy Oxford Street, his dark eyes fixed on me, is a man I've never seen before. Just before we pass each other, he drills a closed fist into my chest.

The next night I call Paul and tell him what happened. He tries to talk me into coming home. I tell him I'm going to finish my research. Instead of being sympathetic, he's angry.

When I get home, Paul's bullying escalates—quietly, carefully, usually through questions meant to shame and silence me. One day I tell Paul that I don't want to be the only one working this hard on our relationship. If he wants me to stay, he's got to go back into counseling with me. (He's quit going again.) It feels like the only path out of a maze I can't navigate on my own.

"I'm going to try it your way," he says, "I know you think my not going makes a significant difference. I think it makes a minuscule difference."

My therapist recommends a man who specializes in sexual addiction. I wish Paul cared enough to check him out for himself, but, as usual, he puts it off. I know if we're going to see someone, I'll have to arrange it.

Once we're in counseling again, Paul says he's never wanted to make a relationship work more than this one. "I've never loved anyone this much before," he tells our new therapist.

"Please don't leave me," he says to me when we're alone. "I need you." He sounds like he means it. And foolish me, I keep mistaking need for love.

Chapter 18

Our therapy sessions don't change much at home. Paul is almost always in a critical mode. One afternoon, just before we leave for a political fund-raiser for my friend Alex, Paul looks at me and says, "You're wearing that?" This is his new mode of attack. I look down at my trouser suit. Then I look at his turtleneck and well-worn khakis. Gone are the fine wool trousers and jackets, and the compliments—along with the man who used to make an effort.

These days, if I put on lipstick, he tells me he thinks makeup makes women look like whores.

"Are you saying I look like a whore?" I ask.

"That's not what I said," he replies. No amount of arguing shakes his cruel excuse for logic. I rarely dress formally, but when I wear clothes a bit more formal than his, he asks me whom I'm trying to impress. I realize such words have little to do with me, but it's hard not to feel stung.

When we get to the fund-raiser and I introduce them, Alex shakes Paul's hand and says, "You're a lucky man."

"Well, it's been interesting," Paul says.

If he was hitting me, I know what I'd do. I have a strict rule about that. But cruel words are what I come from, so my tolerance is high. The way people in my family deflect responsibility for them is by accusing the target of being too sensitive—that, or denying that hurtful words were said in the first place. I decide to speak up anyway. The next day I tell Paul I'm not interested in being around him when he treats me this way. His response surprises me.

"I'm sorry to hear that," he says.

I ask him if he knows why he criticizes me the way he does. He responds with a question: "Why do you want to be in a relationship with me when you're so discontent most of the time?" This is the most rational moment of the next hour. It's hard to say who's angrier. I go upstairs and fall asleep. Being with Paul exhausts me. At some point, Paul comes in and kisses the back of my neck.

That night, Paul wakens me from a deep sleep. I feel his arousal and turn toward him. It's the most satisfying sex I've had since our first night together.

In the morning, he wraps me in his customary python embrace, and says, "That was nice. Was it for you?"

"Yes," I say. I don't ask who he thought I was, but I know it was good for me only because I wasn't fully conscious and because I was thinking of Jim. I keep this to myself—whether I believe Paul deserves such kindness or not.

Paul gives a party for a friend who's just gotten tenure. It's someone Paul thinks I should like—you're both strong feminists, he'd said—but, in fact, she makes me nervous. When she's around Paul, she acts as if she's an ex-wife and I stole him away. As usual, neither Paul nor I have to do anything to prepare. The house is cleaned once a week, and the dinner is being catered. By the time the party is over, I've pieced together one fact: Paul has a long list of women whom he supports: some with monthly checks; some with one-time payments, for a house, for education; some with steady streams of expensive gifts; others with occasional but regular gifts of money.

I ask him about this after everyone leaves.

"I've got so much money, and most people don't have as much as they need," he says.

"Yes, but doesn't it ever occur to you that it doesn't make for healthy relationships?"

"What're you talking about?"

"I'm talking about the fact that when people are financially dependent on you, they aren't necessarily going to be honest with you. They'll do and say what keeps you happy so the money keeps coming in. They'll use you."

"That's a horrible way to look at it," he says.

"It's realistic," I say. "I've seen some of your friends drop huge hints about what they'd like to have but can't afford. I've watched you buy what they yearn for. I just didn't put it all together before."

I mention the name of one of his girlfriends. "Look, Paul, I know you support her. Your aunt told me about it. And I hear how you talk about her, how you couldn't get her to leave when you told her your relationship was over. You make her sound pathetic."

"It's sad to be so cynical," he says, as he walks out of the room.

When Paul goes upstairs, I start to carry dishes into the kitchen. Save it for the cleaning woman, he'd said as usual, but I can't stand leaving a mess for a couple of days. As I'm stacking dishes in the dishwasher, I hear a pounding upstairs. I call up to him to see if something's wrong. He doesn't answer, but the regular methodic beat persists. When I go up to the bedroom to see what's going on, I find him standing against the wall near my side of the bed. He's leaning his forehead on the wall and pounding his right fist into the wall, over and over.

"What's wrong?" I say, standing near the top of the stairs.

"I'm doing this so I don't do something I'll regret," he says, not looking at me.

I ask again: "What's wrong?" It's as if I'm looking at a stranger who wants to hurt himself. Or, it occurs to me, someone else. Everything inside me feels as if it has stopped, except my heart, which is beating in overtime as if to make up for the rest. I feel heavy, laden with new insight. Don't forget you know this, I tell myself.

"I can't stand it," he says.

"What?" I ask, still confused.

"The way you look at things," he says. "I hate it."

Suddenly the room itself looks alien, as if I've stumbled through the wrong door. I go into my bathroom, wash my face, put on my pajamas. I bring clean clothes with me when I go back downstairs to sleep in the guest room.

Chapter 19

In London, where we've decided to spend Paul's spring break, he drops another bombshell. We've spent an unpleasant afternoon: mostly he picks apart the clothing and makeup of women we pass. Foolishly, I argue with him, as if it would make a difference. Finally, I go off on my own. That night when we get into bed, he says, as if we're continuing a conversation we'd started earlier, "I'm going to quit taking my antidepressants."

This is the first I've heard he's taking them. I can hardly believe he's kept something so significant from me. "Why haven't you told me about this before?" I ask.

"It doesn't concern you," he says.

I can't make sense of someone who thinks like he does. "Of course it does, Paul."

"No, it doesn't. I've cut back, but now I'm going off them completely."

"Have you talked to whoever prescribed them?" I ask, trying to put a little logic into the conversation.

"I'll tell him eventually, but I don't like how it affects me sexually."

My first thought is that this is the least of his troubles, but I stay silent as I stroke his arm, for comfort—mine as well as his. That's what's in my heart, but he misreads my intentions.

"Go ahead," he says in a cool, flat voice. "Please yourself. I'll watch if you want."

This instant fractures the time line of my life. The next thing I do must be the start of something different. As if instinctively to conserve energy, my body temperature drops. I'm colder than I've ever felt before, colder even than the night I've always thought of as the worst of my life: the night of my father's final drinking binge, when he turned his fists on me. I get out of bed, walk to the bathroom, and sit on the edge of the white claw-footed tub. I'm beyond tears. My heart has finally slammed itself shut against Paul.

I've become one of those women people see on television and don't believe. Our past prepares us for the job and keeps us at our post; we're raised to talk ourselves out of the danger we're in. But somewhere, deep down inside, we know. What we don't know at the time is that we make our monsters even more dangerous each time we give them a pass.

When I come back into the room, I pack my bag and call the reception desk. There are no available rooms in the hotel. Luckily there's a second bed in our room, so I get into that one. When he calls out to me several times during the night, I register his voice as if it has nothing to do with me. Finally, I'm strong enough to answer his cruelty with silence of my own. Besides, I have only one thing to say: I'm through. The longer I'm with Paul, the dirtier I feel. If I don't leave him soon, I may not be able to find my way ahead.

I'm not ready to talk about leaving, though. I don't know what he'll do when I tell him.

* * *

It's as if he knows anyway. "Does it bother you that I have a gun in the house?" he asks in bed one night after we've been home for a few days.

I'd fallen into that pre-sleep state when uttering one more word feels like a mammoth effort. Suddenly I'm wide awake—and afraid. I force myself to ask, in a measured voice, as if I were asking the time: "What kind of gun?"

A pistol, he says.

"Where is it?"

"In my closet."

"Your bedroom closet?"

Yes, he answers.

"Where are the bullets?" I ask.

Downstairs, he says.

Okay, I tell myself, no need to be afraid. The gun's upstairs. The bullets are downstairs. Then I say, "Yes, it bothers me. I don't want a pistol in the house. Why do you have one?"

"I like having one. It makes me feel safer."

"Well, I wish you didn't have one." Then I recount all the reasons it isn't, in fact, safer at all. "I'd prefer if you'd get rid of it," I say.

"I'm going to keep it," he says, as if there wasn't ever a chance he'd change his mind.

A few days later, I tell Paul that if I'm going to live with a gun, I'm going to learn how to shoot it. "So do you want to take me to a shooting range?" I ask. "Or should I ask someone else?"

He acts pleased by the request. That Saturday afternoon, we drive to a shooting range several miles away. From the outside, it looks like all the other storefronts in this suburban strip mall—the video store, the pizza place, the liquor store. But inside, it feels different. I suspect we're the only ones here without tattoos.

Before I load the gun, I hold it in my right hand and acclimate myself to its weight. We spend an hour or two taking turns shooting. Paul's surprised at what a good shot I am.

"I took gun safety when I was thirteen," I say. My dad and brothers hunt. I didn't want to be afraid of guns. Actually, I'm not afraid of guns. I'm afraid of people who can't live without them.

We leave the store and on the way home pick up videos and takeout Italian. Just as if we've done something ordinary, like going to a movie.

Every few nights after that, always when we're in bed, in the dark, ready for sleep, Paul asks, as if for the first time: "Does it bother you that I have a gun in the house?"

My answer is always the same. Our conversation is always the same. Nothing comes of it, except that it makes me step up my plans to leave. I start looking for an apartment.

Chapter 20

The gun comes up again, on a Friday afternoon. When I arrive home from a meeting, Paul isn't alone. He's in the living room, playing pool with his friend, Brad. I've met him before, and I don't like him. Short, he has the body of an ex-military man, taut, tightly wound. His eyes are alert, but not kind. He's the friend Paul called when Greenwood tried to kidnap my sister's children. "One of the best computer hackers in the country," I remember Paul saying.

I take Paul aside and remind him I have to prepare for a speech I'm giving tonight at a literary event. Voices carry throughout the nearly doorless house. It's difficult to get away from the sounds of a pool game in the middle of the living room.

"He called after you left," Paul says, "so I invited him for a game of pool. He needs to borrow my gun."

"Gun?" I ask.

"Yes," he says, "my pistol."

"What does he need a pistol for?"

A job, Paul says.

"Job?" I ask.

Yes, a job, Paul says again.

I force myself to quit repeating his words. "What kind of job requires a gun?" I ask.

"He used to be in the military. Now he does lots of security work."

"Why doesn't he have his own gun?"

"He needs one that can't be traced."

"Traced?"

I can see that my questions won't get me any closer to the truth. Besides, I don't need to know the specifics to understand that I won't like them. I go upstairs, into my bathroom, which *does* have a door. And a lock. I use it. Sitting on the floor with my back against the locked door, I close my eyes and go over the words I'll say tonight.

A while later, I hear Paul come into the bedroom and enter his closet. For the gun, I imagine. He calls my name. I don't answer. He calls my name again. Again, I keep silent.

I have plenty to say, but nothing that will make any difference.

A few hours later, after my speech, I find my seat in the auditorium. When I sit down next to Paul, he doesn't look my way. "I was nervous," I whisper. "I'm afraid it got in the way of my words."

He doesn't bother to look at me as he says, "I've seen you do better, lots better."

Leave him. This minute. The words are coursing through my brain. I've become what I never meant to be: someone else's emotional punching bag. I need to ignore the inner voice telling me that leaving is cowardly, because I need to figure out how to leave.

Later that evening, after the program, a number of guests

compliment me on my presentation. A retired English teacher says to me: "Your words, as always, are beautiful and fresh. I can always depend on you to give me a new way to look at things." I feel grateful for this compliment—pathetic, really, for needing it so much.

I believe in living an engaged life, and the risk is always the same: the possibility of failing, making mistakes, saying something silly, doing less than we intend. I don't want to be with a man who sits back and watches the rest of us, the better to feel superior.

That night, in bed, I hug my edge of the king-size bed and try to do with the evening's kind words what I told the audience to do with a short story: hold them in the palm of my hand and treasure them like a fine silver ball. I'm so immersed in my own thoughts, I don't see his pillow talk coming: "Does it scare you that I keep a gun in the house?"

I assume he gave the gun to Brad and tell him so.

"I know, but he won't have it for long. I just wanted to remind you."

I say nothing. But, just as he intends, I'm afraid.

And then comes one of his other mantras: "I love you more than I've ever loved anyone." These days he usually says this only in the dark.

The part of me that once fed on these declarations of love is shrinking to almost nothing. I love words. I honor them. I think they mean something, and I mean the ones I say. But I must practice ignoring his words. I must pay attention only to his actions. That will be my way out.

Chapter 21

I keep up the charade of being a couple. I don't want Paul to know about my plans yet, and I feel safest pretending nothing has changed. One Sunday afternoon, we go to a mall to run errands. I watch his eyes widen at the sight of young girls in tank tops. He comments on their revealing clothing, their bodies. "Don't you love to see girls just as they're beginning to get breasts," he says.

I feel sick. "Do you have any idea what you're doing?" I ask him.

"What?" he says, as if he can't imagine my meaning.

"They're children, Paul," I say. "I bet they're not even thirteen."

But look how developed they are, he says. "It's exciting to think they're just coming into themselves sexually."

"That's sick," I say, and walk away.

The next weekend, he goes to Chicago for a few days. "For a break," he says. "I need some time alone. I'll be back Monday night." He leaves late on a Friday afternoon. I'm relieved to see him go. It gives me more time to look for an apartment. He calls several times during his seven-hour drive. "I don't know where

I'm staying yet, and it'll be late when I get there. I'll call you in the morning so you know where I am."

But I don't hear from him until late Sunday afternoon when he calls from his car. He's on his way home.

"You're coming home a day early," I say.

"Yes," he says, "I've had what I needed. And I miss you."

"You said you'd call," I say.

He says he forgot. "Don't wait up," he adds.

I know in my gut that he's not been alone. But this time I'm not going to ask. It's a moot point. Besides, I wouldn't expect the truth. Several hours later I hear him come into the house and up the stairs. I feign sleep. I'm lying on my side, with my back toward him. He gets in bed and edges toward me. He wraps his arms around me. Tight. I roll my shoulder and move away from him, as if he's disturbed my sleep. He moves closer and wraps his arms around me again.

"Leave me alone," I say. I bring the cotton comforter up to my mouth to keep from gagging. Addiction freshly fed (I know it the way your gut won't let you avoid the truth), he turns to me to make him feel human. I won't do it anymore.

"What's wrong?" He sounds more guilty than confused.

"Just leave me alone," I repeat.

There's only one thing to talk about, and I won't be doing it in bed.

I pick the first night that feels like spring to give Paul my conditions, which I see as the easiest way to bring this thing to an end. I'm counting on him saying no. "If I stay and give this another chance, here's what you've got to do," I say as we stand in the kitchen. "You work on recovering from your sexual addiction, go back to the therapist with me, and tell him everything. And I mean *everything*," I say, thinking of his meds, his trip to Chicago, and the girls in the mall. I tell him he's spiraling downward, and I'm not going down with him.

Any further down than I've already come, I think.

I'm standing in front of the screen door that leads to the patio, to let the spring breeze warm me. He's leaning against the desk in front of his cubbyholes.

"This is sudden," he says.

That actually makes me laugh. "Not really," I tell him, and add that, as far as I can tell, we've been heading here for months. "I've been looking for an apartment since we got back from London."

"Without *telling* me?" he says in a controlled but angry voice.

"Oh, that's rich," I say, "like I'm the one keeping secrets."

"The only thing wrong here is that I'm not attracted to you anymore. You want it to be about *me*. You think there's something wrong with *me*," he says, and crosses his arms.

"There *is* something wrong with you, Paul," I say. "You're a sex addict. You told me so yourself. You were recovering before, and now you're not. So I'm leaving." As soon as I pack my things and find an apartment, I tell him.

Quick as that: ultimatum, response, conclusion. I thought it would be protracted and am relieved it's not.

"You don't have to leave right away," he says. "Stay for the summer."

"Thanks anyway," I say.

"I'll leave the house if you want," he counters. "You can spend the summer here."

"No," I say, not even tempted by the idea.

"I *want* to leave for a while," he insists. "I *want* you to stay as long as you'd like."

"Don't be absurd," I say. "I'm not about to become one of those women you love to talk about." I can see him lying about me to his friends, even as he's trying to talk me into staying, so I mimic what I've heard him say: "You wouldn't believe how long it took me to get her to leave. It took me *months* to get her out of my house."

The skin on the outside edge of my right eye is twitching. Over the past few months, I've developed a tic that surfaces only when I'm talking to Paul. Tic. Talk. Tic. Talk. It's twitching like hell right now. I want to run out of the room, the house, actually.

"I can't wait to get out of this house," I say.

"I'm going for a walk," he says.

I gather what little dignity I still feel and walk out of the room. The minute my back is turned to him, I begin to slip from the moorings of this place I tried to make my home.

Chapter 22

If this were my house, I think as I walk upstairs, I'd call and have the locks changed, just before I opened the upstairs window and threw all his things into the front yard. Well, actually, I wouldn't do that, but if I were writing a movie about a character with a little less control of herself than I have, that's what I'd have her do. It would give me such pleasure. It gives me pleasure just thinking about it.

But this isn't my house, so instead I clear my things out of our bedroom and move into the guest bedroom, which is where my own bedroom furniture is. When I go into my bathroom to bring what I need downstairs, I see blown-up photos of me laid out on the vanity that stretches across one wall. I can't recall his taking these pictures of me. He must have put them here when he came upstairs before he left for his walk. I rip them up and leave them on the cool marble countertop.

Packing up my bedroom was easy. Packing up my office will take longer. Finding a place to live and work while keeping up my business will be more complicated still. Before I go to bed, I bring empty boxes up from the basement and put them in the library so I can start packing my books first thing the next morning.

By the time Paul comes home, I am already in bed in the guest bedroom and the door is closed. I hear him go upstairs. I guess what he's doing: checking closets and drawers to see if I've really done it. In a minute, he's at the bedroom door.

"Can I come in?"

In a voice as unfriendly as I feel, I ask what he wants.

"To talk."

"I'm all talked out," I say.

He opens the door anyway and asks why I'm sleeping down here. I tell him I won't sleep with him while he shops for my replacement. I know that's what he'll do now, if he hasn't started already. When he sees I really mean to stay here, he tries to talk me back upstairs.

"No," I say. "Leave me alone."

"You had no right to rip up those photos. It was a mean thing to do."

"You don't really want to get me started on the subject of mean," I say, "because I won't be talking about me, and you won't come out ahead."

I must look as mad as I feel. He slams the door shut as he leaves.

When I call my parents the next day to tell them I'm leaving Paul, my father has only one question: "Are you safe?" he asks.

"Yes," I say. "It's not that kind of abuse."

"Are you sure you're safe?" he asks. "Because if you're not sure, just walk out. Right now. There's nothing that can't be left behind." I'm grateful for my father's clearheaded advice, but I don't miss the irony that the man giving it is the same man who conditioned me for just such a man as Paul.

I feel curiously comforted by my father's reinforcing, without question, that I'm right to leave. He could remind me that he saw this coming, but he doesn't. When he's sober, things are not complicated for my father. He doesn't say it, but I know he thinks the error was in being with Paul, not leaving him, but I also know he believes it's better to make a mistake in love than not to love in the first place.

My father's not the only one wiser than I am about the danger to which I've become accustomed. My friend Phil comes over one day and hands me a key to his house. "If you don't feel safe, come to our house—any time of the day or night," he says. "Don't worry about calling. Just come over."

Chapter 23

I'm teaching at the college and I've got deadlines for clients, so it takes me a while to organize and pack all my things. Plus, I'm chairing a nonprofit board and managing a significant staff turnover. The apartment I've found—in the neighborhood close to downtown that Paul says he *hates*—isn't available until September 1, so I'll be staying at a friend's until then. Since she spends most of her summer at a house a couple hours out of town, I'll be mostly on my own.

Not being able to go directly to my new apartment complicates the moving. I have no intention of coming back to this house once I leave in a few weeks, so friends have generously offered to come back here on the day the movers come for the furniture and heavy boxes I'll leave behind. In the meantime, I'm taking with me whatever I can manage to fit into my car and a friend's truck.

Nearly every night when he comes home, Paul knocks on the door of the guest room and tries to draw me back upstairs. "Come upstairs," he says. "Please." Even though I feel the physical withdrawal you do after you've been with someone night after night, I'm not even tempted. I miss the heat of bare skin against my back, but I don't miss sleeping with *him*.

Some nights I don't answer. Some nights I weaken and let him come into the room for a few minutes. These nights follow the same pattern. Paul kneels by the side of my bed and lays his head on my chest. "Please, come upstairs," he says.

My answer is always the same. After a few minutes, I ask him to leave me alone.

Stopping the habit of love takes time. It takes practice, even when, in the end, the reasons to hate more than smother any original reason to love. Regret gnaws at me. Will I ever be able to pick out memories I can bear to keep?

Many days, when we cross paths in the house, he continues to try to persuade me to stay in the house for the summer, but I see through him. He wants to be the good guy who didn't rush me out of the house, and he wants *me* to join the women who take his money so they can't despise him without despising themselves.

Not me, I think. Not me.

The sound of digging in the front yard wakes me up a few days before I leave. I walk barefoot across the cool, white tiles and open the front door. A woman who looks to be in her forties, with each year having traveled across her face more than once, is digging up grass on either side of the entry. She's got on short white cut-offs and a bright turquoise tank top. Her hair is pulled through the back opening of a baseball cap.

"Can I help you?" I ask.

"I'm just doing some planting," she says, "as a way of saying thank you to Paul." She introduces herself. I recognize the name. It's a woman Paul dated when he was in his twenties.

"Does he know you're doing this?" I ask.

"Oh yes," she says. "Wanna help?"

"No, I don't think so."

She asks if I'm still planning on leaving.

"Of course," I say.

"That's too bad," she says. "He really loves you. And you're good for him."

I ignore this. "What're you thanking him for?"

Paul's given her another loan, she explains, so she wants to do something to thank him. "He's already paying the mortgage on my house," she adds. "I probably would have lost my kids a long time ago if it weren't for him."

I ignore the last part. "A loan?"

"Well, you know how his loans are," she laughs. "You don't ever have to pay them back."

No, actually, I don't know, I tell her.

"Oh, that's right. You won't take his money," she says. I know from Paul that she's been on and off drugs for years. Whenever he talks about her, he makes her sound pathetic—dependent and unceasingly grateful.

With Paul, money is the tie that binds. Taking money makes it hard to forget a person, and I plan to forget Paul as much as a person's able.

Later that day Paul tells me about a woman he wants to meet. He tells me she works at a women's health clinic downtown. He wants me to offer a large donation in his name and arrange it in the context of a lunch, for the three of us. I can see where this is going.

"I won't pimp for you," I say.

He's so angry he stutters as he starts to protest.

"Save it," I say, "for someone who still believes you."

Chapter 24

It's hot the day I leave Paul's house, but the tile floors and shaded windows keep the house uncomfortably cool. I wear socks and a sweater as I wander from room to room. For all my efforts to make this a home, it remained just a house—his house, with dark corners and not enough doors.

In the front hall, I line up boxes and duffels full of clothes, books, and files—the things I'll take with me. The rest—mostly furniture—is in the spare room waiting for the movers. Paul's offered to help, all along, but I don't want a thing from him, except to leave me alone and stay out of the house my last day here. He acts hurt: just one more way he annoys me these days. As if *he* were the injured party.

I tell him I'll be gone by eight. I really plan on leaving by seven, to avoid the good-bye part of the day. I make plans to meet friends for a seven-thirty movie. Just before I leave, I go back to the upstairs bedroom for one more check. For the last time, I sit on the blue duvet cover I bought after I'd moved in. Paul has told me to take it, but I don't want a reminder of anything we shared.

Next to his side of the bed is a newly framed photograph—of me. Of all the photos he's taken, this is the first he's ever framed. Until this moment, it was my favorite. Right after I'd moved into his house, we drove north and stayed in the woods for a weekend. Through a mass of maple leaves behind me, the late-afternoon sun casts a shadow on the woven scarf covering my head. But mostly what the camera captures is the soft look of love.

This is just like him: yearning for something unattainable instead of appreciating what was staring him in the face when he could have done something to preserve it.

I open the built-in cupboard by my side of the bed. I'd cleared it out the night I moved into the guest room, but I want to make sure I've gotten everything. I find a sheaf of typewritten sheets I thought I'd already packed. Under the pile are handwritten pages, in Paul's hand. I know I shouldn't read them. I do anyway. They're notes about the woman he was with before me, the woman who left him after he told her he was with a prostitute in Thailand.

She and I form a select group: women who've left him. With great precision, the pages detail his early dates with her. He'd hold himself back sexually, and then, as soon as he left her, he'd visit a prostitute. Oddly enough, this is one of the most reassuring things I could possibly find on my way out of this house. As much as it disgusts me, it makes me realize my final instincts about him are right. I also find notes about dates he's had in the last month. Each one reads like a card for a research paper: name, date, details, conclusion. I know I shouldn't read these either, but I can't stop myself. The first one in the stack is dated the week after we agreed to split. After the name are the following details: "34, blond hair, long legs, daughter 12 years old. Call again."

Behind the stack of papers is a black leather belt. Just like the one used to strangle his father.

At the foot of the bed is an open cabinet. The television sits on top. On the large open shelf below are videotapes. Mine are all *Masterpiece Theatre* and *Mystery*. I assume his are all *Nova*.

I slip one of the unmarked tapes into the VCR to see if it's mine. At least that's what I tell myself. The truth is, the papers I've just found make me wonder what else I missed in our bedroom. But nothing I suspect prepares me for what I see: Paul's father having sex with a young man. My hands shake as I quickly eject the tape and shove it back in place. I assume the other unmarked tapes are more of the same. And he kept them, at his bedside—at *our* bedside.

This revelation pales in light of what comes next. One of the tapes in the back row is marked "police property." It has to be the tape Jen told me about: the tape of his father's murder. I don't even touch this one. I walk out of the bedroom and for the last time pass the life-size painting of his father hanging in our bedroom. Good riddance, I think, and don't feel a bit badly for it.

At ten to seven, as I put the last duffel in my car, I see Paul's car coming around the corner. It's more than an hour earlier than he was supposed to return. I hurry into the driver's seat and turn the key in the ignition. Paul parks next to my car and walks over. He acts surprised by the packed car, as if this wasn't the point of the whole day, as if my leaving wasn't the point of the whole month. "Where are you going?" he asks.

He can't stand that I'm eager to get out of his life. "You know I'm leaving," I say.

"But I don't want you to leave tonight," he says.

"I have to go," I say. "I'm meeting friends at a movie."

What friends, he asks.

It doesn't matter, I tell him.

Tell me if it doesn't matter, he presses.

"No."

"Which theater?"

"It doesn't matter," I say again. I know he doesn't like this. But my life is no longer any of his business. I don't want him to know I'm going to my older brother's house; his family is out of town for several days. Then I'm moving to my friend's house for the rest of the summer.

As if to keep my car in place, Paul hooks his fingers over the car window, which is rolled nearly all the way down. Crouching down on his haunches, he drapes his tanned and muscled arms over the door. He smiles. "Please come back tonight," he says, as he leans his head into the open window. "It's too late to go anywhere else."

I shake my head. This show of emotion is only that—a show. "Not tonight," I say. "Not any night."

He reaches for my face.

I pull away. He settles for stroking my bare arm as he says, "Please stay tonight." His fingers reach for my face again. "You'll never forget me," he says.

I pull farther away from him.

"I'll always be with you," he says, as his fingers trail across the back of my hand.

I shudder and work to keep my face from displaying any feeling. I close the window, lock the doors, and drive away.

PART II

1994–2003

Chapter 25

I was born in a Great Lake city and brought home to a Mississippi river town, both bodies of water left in the mineral-rich wake of a glacier. When we exchanged our small town for an urban life, we moved to a city on the same river. Only twice in my forty-four years have I lived more than a handful of miles from the Mississippi, so I feel most at home when its waters are as close as my back door.

My friend's house is a short walk from the river and just down the street from the college. Being in a neighborhood that feels familiar on two accounts eases the strain of not being able to settle somewhere in one clean move. That, however, is minor compared to how leaving Paul feels: as if I'm saving my life.

It's an old hodgepodge house devoted to function rather than aesthetics. Before I go to bed the first night, I unpack enough to make me feel at home. I set a white ceramic mug and a navy-blue tin of Earl Grey tea on the kitchen counter. I unpack a few books and stack them on the worn oak table I'll use as a desk. In an upstairs bedroom, I move a lamp to the bedside, and on the tall dresser I prop my painting of swans in flight.

From the day I leave Paul, I avoid anywhere I'm likely to run into him, which even in a big city isn't as easy as it sounds. Nearly every time I walk into one of my regular hangouts, I hear he's been there recently—even the places he didn't frequent when we were together. One night I go to meet a friend for dinner at one of my old standbys, Niko's, and take a seat at my favorite table, midway down the room.

"I'm sorry about your breakup," Niko says when he comes out to greet me. He carries the look of attentive restaurant owner well: dark slacks and a crisp white shirt that's open at the neck. Mostly it's his clear, dark eyes that welcome you.

"Who told you about it?" I ask.

"Paul," he says. "He comes in a couple of times a week these days. In fact, he was just in last night. He gave a party for friends." He pauses. "Your break is amicable?" It's a raised-eyebrow question.

"Says who?" I ask.

"Paul," he says. "But I wondered."

"He's not the man I thought he was," is all I say, and add the detail that Paul seems to turn up at all my favorite places these days.

"I thought it was funny," he says. "He was never here unless you came in together. Now he's here all the time." He tells me not to let that stop me from coming.

This scene is repeated many times, and, of course, Paul *does* stop me from going to my favorite places. At least for a while, he wins the point. I want to spare myself any connection, even in passing. This must be his way of punishing me for not stepping into his long line of women who are left beholden and grateful.

In spite of my plan to avoid him, he's in my life plenty anyway. He calls several times a day. He talks about trying again, doing better, making it work. I've decided he wants to get me back for one reason only: so he can rearrange the break in order to cast himself as the wonderful and generous one leaving. That, of course, would make me the one being left. Never mind that it would rewrite history: I refuse to give him such satisfaction.

For a while, I quit answering the phone altogether. That's when the drive-bys start. I keep well away from the window and park my car behind the house so he can't see whether or not I'm home. Every once in a while, he comes to the door. I make him stay on the porch as he tells me why he's come. For nothing, really.

Mostly he just circles the block, often calling from his car phone as he does. One day he calls twelve times. The lack of restraint he's showing is so opposed to his usual cool control. This actually gives me a bit of pleasure.

Chapter 26

I drive to my parents' home to get out of town for a while. When I walk through the front door, my father and Liz's youngest daughter are in the living room. "Run to your aunt," my father tells the eager toddler. "She needs a hug." My niece shrieks in delight, runs across the room, and flings herself at me.

My father's more sober hug follows. "Your heart is broken," he says. "You loved him, so your heart is broken." He pauses as if to consider. "For a year," he adds, "your heart will be broken, for about a year." He hugs me again. "And then you'll be fine, because you're a survivor. You've always been a survivor." Wondering what story from his past informs him of this one-year term for heartbreak, I am rocked by such tenderness from my proud old Irish father.

It's more complicated for my mother, whose guiding rule is simple: if you love someone, you forgive everything. The disease of alcoholism swamps us from all sides, from both my mother's and my father's families. Since my mother never actually left anyone (though she occasionally *planned* an exit for us), she was practiced in forgiveness. Her mantra could have been: one more chance.

I didn't share her hyper-optimism. By the time I was in my

late twenties, I'd given up hope for my father's recovery. I'd given up hope that my mother would see that in not standing up to her husband she failed to protect her children. I quit expecting I would love my father again and would forgive my mother. But my parents each did their part—as did I—and proved me wrong to lose hope. That's when I learned to believe in—and practice—redemption and atonement. Redemption not as in a miracle, but as in the relentless emotional work that results in a person's taking responsibility for a life, day after day, the kind of effort that transforms a spirit.

I wonder if my leaving Paul feels like a betrayal of all my mother stands for. Perhaps she thinks I didn't try hard enough to make it work. She makes me feel welcome at home, as she always does, but a few days into my visit she says, "You're not the only one who's ever gone through a breakup."

"I know," I say. "Why would you *say* such a thing?" I haven't been moping around. I feel like a wreck inside, but I work hard to act cheerful.

She won't explain what she means. She could feel foolish to have liked him so much. We all do. But whatever her initial feelings were, before long she adjusts her view.

Once I'm at my parents' house, Paul turns his obsessive dialing on them. I avoid answering the phone, which leaves it to my mother to tell him I don't want to talk to him.

"Do I need to get on the phone and take care of this?" asks my father one day after yet another call.

"No, I can handle it," she replies. And she does. She doesn't want to talk to you anymore, Paul, my mother tells him the next time he calls, and neither do we. She finishes firmly, not easy for her: "Quit calling here."

His calls to their house stop, but as soon as I'm back in the city at my friend's house, they begin again.

One afternoon Paul leaves a message on my answering machine begging me not to give up on him. He wants to figure out a way for us to get back together. "I want to become the kind of person who's able to be in a loving relationship," he says. "I know I'm not that kind of person now." No shit, I think, when I hear his voice on the message.

Maybe he *can* change. My father had. I'd given up on *him*, and I'd been wrong to. I hate thinking I've picked such a maniac. Yet I've promised myself I won't go back, no matter what.

I'm strong, but I'm not perfect, so three times—exactly—I give in and meet Paul, always in public, and never for long. I regret it each time.

After that, whenever I feel weak, I remember the advice of the last therapist we saw together. I'd called him right after I left Paul. He fooled me, the therapist said. He told me that he believed Paul when he said he was prepared to do anything to make our relationship work. The therapist's parting words were: "He'll want you back. Don't do it."

Chapter 27

Some people are oblivious to their surroundings. They can live anywhere. I'm not one of them. It's not a question of size. It's about the aesthetics, the charm of a place, inside and out. The renovated apartment I'm leasing occupies the third floor (the top floor) and part of the second floor in a large house on the edge of downtown. I especially like the peaked ceilings that create cozy nooks in every room: My nieces and nephews will love these ready-made forts. A couple of blocks from a city lake, it's also within easy reach of a theater and art complex, several coffeehouses, and a multiscreen movie theater. I can't believe my luck in finding it.

As soon as the movers and the friends who've helped me have left, I pick my way through cardboard boxes and start to nest: filling bookcases, plugging in lamps, and setting up my desk and enough of the kitchen to have breakfast the next morning. I set out the electric kettle, the toaster, and my well-seasoned Sadler teapot. Finally, I hang three paintings, the swans first. I plan to live here for a long time.

When I arrange my books in my new place, I put the ones I regularly reread in one bookcase and arrange them alphabetically by author—Brookner, Byatt, Cather, Dickinson, Durrell, Eliot, Fairstein, Fyfield, Grennan, Hardy, Jouve, Leon, Oliver, Sanford, Sayers, White. When I lived with Paul, I categorized my books by genre, but I don't do that when I live alone. I must have done it there to put order to a fragment of my life.

Paul takes to calling me late at night. It's before the days of caller ID, so I don't know who's calling until I pick up. One night I'm sitting at my desk, facing away from the windows overlooking the backyard and garden. I'm trying to make sense of notes I'd taken when I was in London last year. The ringing phone startles me. I look at my watch. It's nearly midnight.

"It's me," he says, in the soft seductive voice he used on the phone when we were first dating. "You're still working, aren't you?"

I look outside to see if anyone could be seeing in. "What do you want?" I ask.

"Just to say hi," he says, and he goes on, without a pause, to tell me he's looking at my picture right now. "I keep it at my bedside so I won't ever forget you," he says.

Silence on my part. I don't know what to say. He fills in the gap: "I saw you this afternoon." And as if he's briefing the next shift of a surveillance team, he tells me exactly where and when.

I slam down the phone. The next time I answer and it's him, I ask if he knows where I live. He laughs. I know exactly where you live, he says. Besides, he adds, I can see you whenever I want. From now on, it doesn't really matter if I answer or not. I get his message either way.

This annoys me more than it frightens me. I haven't seen *him* in weeks. I call friends and family and ask them not to tell him anything about me. My sister Liz tells me he's called and he's told her that he knows where I live and that he sees me often.

One day I come home to find a box outside the door to my

apartment. Inside are things I'd thrown away when I left Paul's: papers torn in half, a black leather glove with a hole in the middle finger, an empty print cartridge, impossible to mistake as anything other than garbage.

A few weeks after the move, I realize I can't recall the last time I received any personal mail. I seem to get nothing but junk mail. When I call the post office, the man who takes the call tells me the mail-forwarding order I filed in June was canceled. I tell them I didn't cancel the order. He insists my signature is on the form. I tell him it isn't my signature. I drive to the post office and resubmit a mail-forwarding request.

The next time Paul calls, I ask him if he knows anything about it.

"Your mail is here," he says in a taunting voice. It's been piling up for weeks. "If you want it, pick it up."

"This one's a federal crime," I say. "I want my mail and I want it *now.* I want it left outside my door *tomorrow.*" The next day I come home to several weeks' worth of mail on my dining room table. He knows a mutual friend has a key to my apartment. He had her make the delivery. She thinks she's doing me a favor. I explain that I don't blame her but that I will feel most comfortable if she doesn't have a key. We've become friends, but she's working part time for Paul these days, and I don't trust that he wouldn't "borrow" my key and make a copy of it.

When he calls a few nights later, *I* have something to say. He starts out by telling me how excited he is to be in medical school. As if I'd still care. "I wouldn't be there if it weren't for you," he says. Of course, by now I'm thinking how much harm he can cause as a doctor.

"Do you think I still love you?" I ask.

"I know you do," he says.

"Well, you're wrong. Don't ever call me again. And if you ever see me in the street, you'd better turn around and walk the other way. Because if you come up to me and talk to me, I'll spit in your face."

"You wouldn't dare," he says.

He may be right. I *feel* this way, though I'm not sure I could actually do it. But I keep my voice firm: "Try me." I wait for a few seconds. "I look like a lot nicer person than you do, so everyone would assume you're the asshole."

These are the last words I ever speak to him.

Chapter 28

One of Paul's friends, who ostensibly calls to say hello, mentions that Paul claims to have seen me the day before. This is becoming a familiar refrain; whenever I run into one of his friends, I hear the same thing. This last time, the man goes on to say, was at the video store in my neighborhood, which is tucked into a dead-end street. When I'm there, when I'm anywhere, I constantly scan the street for Paul's white car. I never see him, no matter how carefully I look. The only way he could have seen me is if he was parked nearby, hidden, for no other reason than to watch me. But it can't always be Paul watching me. He can't possibly follow me all the time himself. He's too lazy for that. He must hire people to do his spying for him. This scares the hell out of me.

Just before he hangs up, this friend repeats what they all tell me: Paul is furious I won't stay in touch with him. "He likes to be the one to leave," he says.

One by one, I break contact with anyone I know who has contact with Paul. "For safety," I say, so they'll pay attention. But I really mean privacy. I don't yet know how afraid I should be. All I know is that he's breached all the boundaries of my life.

According to his friends, he even has his new girlfriend dressing like me.

Paul continues to phone my siblings who live out of town, even the one with an unlisted number Paul never had. He often calls them just before I visit them. He tells them where he sees me and fishes for information. They all agree he's having trouble getting over me.

Then Paul's surveillance turns silent. In the evenings, when I pick up the phone, more often than not I'm greeted with silence. I quit answering the phone entirely. But it angers me, because it shuts out everyone else, too.

I'm certain that he'll tire of this. I don't think any of it is worth reporting to the police.

Chapter 29

Jen and I are making an effort to stay in touch, but it's not easy for either of us. She and Doug try to convince me to move back into their building. "It's your home," they say, but I know this isn't an option.

"I want to visit you in your new apartment," Jen says on the phone one day.

A few days later, we sit in my new living room and try to have one of our old conversations. The discussion, however, soon turns to Paul. "I've only see Paul cry twice in his life," she says. "He cried when his father died, and he cried when he came to tell us you left."

I don't know what to say, so I don't say a thing. Everything with Paul is just an act, I remind myself. He knows how much Jen loves me. It wouldn't be good for his image if he didn't seem sad to lose me.

"You're the only woman he's been with he couldn't control with his money," she says.

"It wasn't easy," I say. "You always thought I lived off his money anyway."

"We did at first," she says. "Everyone else had, but now we know better."

"I loved him," I say. Now I feel stupid for it, but I know it's kindest not to say *that* part aloud.

She looks pained. "I know, and I think you may know him better than anyone else. You certainly know him better than we do. He was always so secretive, even when he was young." She pauses, then adds: "I don't think he likes people to know him too well."

We're sitting next to each other on the couch, the same one we sat on the morning she told me about Roger's murder. "Here's a real surprise," she says. "He sold his house. We never thought he'd sell his family's house. He's moving into a downtown apartment. He says he can't stand staying in the house where he lived with you."

This makes me nervous. "Where's his new apartment?" I ask. She gives me the address. I feel sick to my stomach. It's only a few minutes' drive from my neighborhood.

"Do you know about the videotapes Paul has from his father's house?" I ask. "And not just the ones of Roger's murder." I didn't plan on talking to her about this, but it seems right to.

"We tried to talk him out of taking those tapes," she says. "The police wanted to destroy them."

"I found them the day I left," I say. "I didn't know about them before."

She works her mouth as if she's trying to decide something. I'm not prepared for what she says next. She asks if Paul ever told me his father was arrested for molesting two young boys.

"No," I say. I know what this question costs her. I don't want anything to stop her. "He never told me."

The boys were about twelve, she says, and Paul was eighteen.

"He never told me," I repeat, but I tell her that I know about the boys from Thailand who'd lived with Roger years earlier. He'd met them on a vacation and arranged for them to live with him. When they returned to Thailand, they were considered foolish and ungrateful to turn down a chance at a better life. That was the story I'd heard. Now I ask, "Jen, didn't anyone ever wonder about the boys from Thailand?"

No, she says, but now she can see they should have. She's intent on continuing her original story. "Paul went with his mother to pick Roger up at the police station," she continues. "I don't know what it took, but the families didn't press charges."

I think she means money, to silence them. I can't bring myself to ask her if she's considered that Paul might be a victim of incest. I don't know if she's carried her own thinking about her nephew that far.

Well, I tell myself, no secret is sacred now, so I tell her about the visit Paul and I made to the East Coast a year earlier. One night we had dinner with Hal, the man Paul traveled with on photo shoots. Older than Paul by two decades, Hal assumed I knew the truth about their trips. Paul traveled with him as an unpaid assistant and took care of all their expenses. That way Hal made extra cash by putting in for expenses Paul had already covered. No one ever caught on, he explained, laughing.

I didn't join in their laughter. Instead, I pointed out how dishonest it was.

It didn't hurt anyone, said his friend. Besides, he added, they had great times together.

Later, when I was alone with Paul, I said, "You assume no one will be with you if you don't make it financially worth their while, don't you? Is there anything you don't think you can buy with money?"

"Not really." He shook his head. "That's how people are."

"No," I said. "That's how *you* are, and most of the people around you, evidently, but it's not how *everyone* is."

When I tell Jen this story, she says, "I always wondered why he never got credit in the magazine. He told me he didn't want credit, so he always let his friend have it, even when he took the pictures himself. At least that's what he said." She pauses. "See, that's what I mean. You've figured out a lot about him. He won't like that."

When she leaves, I replay the scene she described of Paul going with his mother to pick up his father at the police station. I can see them going into the jail. I can picture his father acting angry, as if *he's* the one who's been wronged. And I imagine money exchanging hands, to guard the successful businessman's reputation. It appears to have worked for him. Up to a point, of course.

Chapter 30

The rocks in Maine's Acadia National Park, wide and flat, look as if they're waiting for someone to slide off them. As I walk along a trail high above the shore, it's easy to see why the Wabanaki Indians called this island Pemetic, "the sloping land."

Coming to Maine is part of my journey back to solitude without loneliness. I consider it a simple matter of recalibration: adjusting the colors of my life so red isn't the only color I see. What angered me most when I left Paul was realizing I'd lost the knack of being content on my own. Before Paul, I'd never lived with anyone. I liked having my own place. I don't miss Paul now. Rather I miss the *pattern* of my previous life—being independent, without being afraid.

I'm visiting a younger brother and his wife, who are renting a house on the coast, near Blue Hill. Just before I arrived, Paul called them; we can't figure out how he got the number, which isn't listed. By now, we're all staying well clear of him. My brother didn't take the call and didn't call back. All week long we feel Paul's shadow: I know where she is, all the time, and I can get to her, and any of you, whenever I want.

We've come to Acadia together, but I split off from my brother and sister-in-law earlier in the afternoon. After walking for a while, I stop and sit on a rock as gray as the sky. Suddenly I feel watched. I look behind me and notice that the hiker who stopped a few yards away moments earlier hasn't moved on. He was behind me all the way from the parking lot, lingering when I slowed to watch a painter catch a scene or the ocean curl toward shore. Now he has stopped and is staring at me. We're the only people on this stretch of trail.

I turn away and focus on a flat line of heavy, steel clouds just above the horizon. But within seconds, I look back at the hiker. He's still there, and this time he speaks to me. "I've been watching you," he says. "You're alone now, aren't you?" His inflection makes it sound more like fact than question. He's wearing tan pants, sturdy hiking boots, and a dark brown fleece jacket. He's young and fit. He looks strong.

I scowl and say, "No."

"Do you mind if I sit with you?" he says, and, as if my answer will have no bearing on his actions, he moves closer and starts taking off his backpack.

"*Yes*, I mind," I say. "Leave me alone." I move as if to rise. "I'll yell for a park ranger if you don't."

"Come on," he says.

"I'm *very* serious," I say, enunciating my words for emphasis.

"Never mind," he says. He puts his backpack over one shoulder and, turning in the direction he'd come, breaks into a run. I can't shake the feeling that he's here at Paul's bidding.

Suddenly, an image flashes through my mind: a body being pushed over the cliff onto an instant, rocky death. The body is mine. I'm not fearful by nature, but months of being watched make me feel vulnerable to unseen danger. Besides, this is bringing back a horrible memory—of a woman who jumped to her death from my apartment building years ago. I'd found her body sprawled in front of my car. What I remember most vividly was that she had on only one shoe, a brightly colored embroidered silk slipper. The other must have fallen off during her descent from the seventeenth-floor balcony, just one floor above my own. Desperate to scrub off whatever had splattered my windshield, I drove through a car wash three times that day.

So I know what a body looks like when it lands after a jump (or a push), and I don't intend to be one. I move back from the edge of the rocks. If someone goes over the cliff today, I think, it won't be me.

I will my body to be heavy, and solid. I will myself to be safe, and sane.

Chapter 31

Wallowing in regret doesn't do a thing except sap your energy, so I point myself toward one goal: re-creating a solitary life. Well, two goals, really. I also want to figure out why I was vulnerable to Paul's charms and manipulations in the first place. One of the first things I did after I left was to schedule weekly sessions with my own therapist. I'm determined to understand what led me to love Paul. I think of life in terms of stories that offer insight. So I look back and begin to tell my therapist the earliest stories I remember that led me to Paul.

The summer before I started school, our family's fifth child, and fourth son, was born. As the third child and only daughter so far, I was sandwiched between pairs of brothers. I can still hear my father's words the day the new baby came home, immediately echoed by my mother's mother. Your mother depends on you to help her with all these boys, they both said. You need to be her number-one helper. Never mind that I had brothers three and four years older. Never mind that I wasn't even in kindergarten. My role in the family was set, and I took it to heart.

If I was hazy about whether my role applied only to my siblings, the death of my grandmother the spring I was twelve cleared that up. I'd heard my parents talk about how badly my grandfather was handling his sorrow, and, as much as I loved my mother's father, I was dreading the monthlong visit they planned for me that summer. However, I didn't know how to say no, so I went.

"I want to die," my grandfather said several times a day. "I pray every night that God will let me die," he'd add, angry every morning that he woke up from his long, drug-induced sleep. All day long I heard about his contempt for a God who didn't answer his new prayer. How could a twelve-year-old talk him out of *that*? I couldn't imagine a way. All I knew was, in that house with shades drawn on every window, it was my job to keep him company in his dark cave of grief. For that matter, it was my job to keep him alive, make sure he didn't take too many sleeping pills. I'd been carefully instructed.

I spent my thirteenth birthday there, a memorable occasion because it marked my first panic attack—the only one I had until I lived with Paul. While my grandfather still slept, I went to mass at the neighborhood church. I dressed carefully, making an effort at happiness: I was wearing the bright floral dress my grandmother sent shortly before she died. Suddenly, just as the priest raised the chalice full of wine, I went cold and weak, as if blood were draining out of my body. I was afraid I'd faint if I didn't get into the fresh air, so I rushed out and sank down on the stone steps outside the sanctuary. I crushed the cotton skirt over my face to hide my sobs. A few hours later, when I told my grandfather about it, all he could manage to say was, "You should have stayed in bed."

That summer was my boot camp. It set me up to feel as if part of my purpose in life was to keep people safe—my grandfather, my mother, my siblings. But I knew this was no way to live. The more I saw of what my mother and grandmother sacrificed—for years at a time—their hearts most of all, the more determined I was to have a different kind of life.

Of course, in choosing Paul, it hadn't worked out that way for me.

Chapter 32

If you grow up surrounded by the deep greens of northern forests and the crystal blues of sparkling lakes, the desert's sand-washed colors seem a pale excuse for nature. It always takes me a few days to adjust my expectations and appreciate the Southwest. I've flown to Arizona to help my parents drive home. They've been here for a few months, and Mom isn't sure Dad's strong enough to make it home. He's not so sure either.

Actually, most of us are surprised Dad has lived as long as he has, given the massive stroke he suffered nearly sixteen years ago, at the age of sixty, just months after he started recovering from alcoholism. He has an enormous will to live, but his health continues to decline.

"If I can only make it to the party . . ." he says to me the first time we're alone after I arrive. He leaves the thought incomplete, but I know what he means. In a few weeks, we're having a celebration for my parents' fiftieth wedding anniversary.

We leave Rio Verde early in the morning, and drive north on Highway 17 until we connect to Interstate 40, which will bring us east through Albuquerque, Amarillo, and Oklahoma City. After that we'll head north again. I like having my parents to myself.

And I'm eager to keep accumulating happy memories with my father. I'm also glad of an excuse to be away from home, to have a break from being watched all the time. I don't get calls from Paul with reported sightings anymore, but often enough I bump into someone he knows and I continue to hear: He saw you yesterday, or last week, or earlier today. But, of course, I hadn't seen him.

As we pass through New Mexico, Dad, in the front passenger seat, points to a field of tall grass the color of burnt wheat. "They've started to replant the native grass," he says.

"Do you know the name of this one?" I ask.

"Nope," he says. "I just know it's the one I like best."

Mom had driven most of the morning, and now she's nodded off. Dad reaches for the visor and puts it down so he can see that she's still sleeping in the backseat. "This will be the last time I see this," he says, "so I find it especially beautiful."

No maybe, no might be, just a statement of fact. They drop me off at home a few days later, and two weeks after our return, we have the party. Dad suffers a massive stroke three days afterward. For exactly seven days he's suspended in a coma. During that week, Paul calls Andy, Liz's husband. Just wanted to say hi, Paul says, but Andy suspects Paul heard about Dad and is fishing for information. Andy doesn't tell him a thing.

On the evening of the seventh day, my father dies, at exactly the moment I stood in Paul's kitchen one year earlier and told him I was leaving. The same day, to the hour, nearly to the second. The coincidence rings like a blessing.

"For a year," my father predicted after I left Paul, "your heart will be broken." And for that year, my father stood witness to the healing of my heart. I don't interpret the timing of his death as a conscious act, yet, on some deeper level, the year he kept vigil over my heart felt like my father's final atonement.

When I go to his hospital room to say my private good-bye

to the body that no longer has a beating heart, I stroke his brow and trace the lines of his face. I have his high cheekbones and his jaw, so strong it announces a face. I also have his ability to see and say a thing straight. I thank him for staying the year, and for the years of amends. I think of the day he said to me, "Your life is so different from mine. I can't picture it. Tell me what you do in an average day so I can see it, from the minute you wake up to the minute you go to sleep." So I told him, about my teaching and writing, about my volunteer work and my friends. When I finished, he said, "No wonder you're a happy person. You have a good life."

During the wake, my brothers take turns looking out the window of the funeral home. I ask one of them what he's looking for. They're checking for Paul's car, he tells me. They think he'll come, and they don't intend to let him in. I'm relieved to know I'm not the only one who sees through Paul's trick of putting forward the sympathetic face.

At the cemetery, my youngest brother recites the Yeats poem that perfectly captures how Dad, who was most at home on a lake shore, imagined heaven: "I will arise and go now, and go to Innisfree, And a small cabin build there . . . And I shall have some peace there, for peace comes dropping slow. . . ."

Chapter 33

Waking up to find my phone dead becomes a regular occurrence, and a mystery the phone company can't seem to solve. One day I pick up the receiver, and, before I dial, I realize someone is on the line. I ask who it is. Three times. Finally, a man answers that our lines must have crossed. I ask where he is. He names the street. That's only a block away, I say, and—suddenly frightened rather than annoyed—I hang up the phone.

Downstairs, I ask to use my landlady's phone to call the phone company and arrange for yet another service appointment. When I go back upstairs, I pick up the phone. It's dead.

On the day they're scheduled to fix it, I wait long past the scheduled time. After a while, I look out the front window and see the company's service truck parked down the block. Thinking the technician is there for another service call, I continue to wait for the doorbell to ring, but I soon realize he's just sitting in his truck. I walk out and ask him why he hasn't come in. He just looks at me and smiles.

I suspect the technician has been paid to do this. I'm meant to look crazy, to scream out my anger at the ordinary-looking man on this bright summer day. Instead, I go back inside the house and close the door to my apartment. Goddamn maniac, I say to myself aloud, but in a quiet and controlled voice. Goddamn maniac. You will *not* win at this. You *will not win*.

Back downstairs, I ask my landlady if I can use her phone. When I call the customer service number, I say, "I've been here all day, and I can see him parked outside. Why hasn't my phone been reconnected?"

The woman asks for my number. I give it to her and wait. She asks me to repeat the number, and when I do, she tells me there's no record of it in the system. She asks for my name, and when I respond, she says that there's no sign I've ever had an account with the company. "Is there another name?" she asks.

No, there's not another goddamn name, I think. I don't say the words aloud because I'm determined to remain calm and reasonable. I *will not* act like the crazy one. I've had an account with you for twenty years, I tell her, and this same number for more than fifteen of those. Whenever I move, I explain, I always transfer the number.

"We have no record of you," she repeats.

When I ask to speak to the service manager, she says it must be a computer error. "This particular thing has never happened before, but that doesn't mean it *can't* happen."

I think back to the day Paul introduced me to his friend Brad, "one of the best computer hackers in the country," the same friend who tracked down the man who tried to kidnap my sister's children. Too quickly, I now realize.

I even begin to wonder if the technician outside rerouted the call I just made. Maybe I wasn't talking to a customer service rep at all. Normally such a thought would seem paranoid. In this instance, it just seems smart to consider the possibilities.

I hang up the phone. I don't know how long I can live like this.

Chapter 34

"I don't want to lose touch," Jen says when she calls one morning. "How are you?"

I mention the phone incident and how threatening it feels that Paul wants me to know how often he sees me, but I don't tell her everything. It doesn't seem fair to burden her, especially since I have no proof.

When she asks me if I'm afraid of Paul, I know that's the true reason for her call. I'm surprised and I wonder what prompted the question, but I don't want to divert her, so I simply say, "Yes. Do you think I should be?"

"Of course I want to say no," she says in a rush. "He's my nephew. I don't want to believe he's capable of hurting anyone." She pauses. "But he's always been so secretive. You know that. And some of his friends"—she takes a breath—"they scare me." She lets the thought drop without further explanation. "He doesn't let many people in, but he let *you* in. You probably know more about him than anyone."

I decide not to ask what precipitated this warning. It's enough that she made the call.

But she's not finished. "If you think there's reason to be afraid of him," she says, "then you're probably right. I don't want to believe it, but I wanted to tell you I think you should trust your instincts."

It's the last time we talk about Paul.

But he's front and center in my life anyway. One day I come home and my door is unlocked and open. I'm *not* careless. I always check to make sure I lock up before I leave, especially these days. I ask my landlady if she needed to go into my apartment for anything. No, she says, I always tell you ahead of time. As we agreed, she adds.

As I walk through my apartment, spread over two floors, I find a bar of soap from the second-floor bathroom on the third-floor kitchen counter. A teaspoon from a kitchen drawer lies on the middle of my bed. A book from my desk is on another kitchen counter. Impossible to mistake as absentminded acts. Some days, the door is open and nothing is out of place. For a while, I put it all down to bad luck, but after months I finally accept that bad luck has nothing to do with what is happening to me, unless you consider that I'm unlucky ever to have met Paul.

One day I wake to find the power out. It's not the first time. It may be possible to have an isolated outage, but by the second and third times (since the power's always on everywhere else in the neighborhood), I know it's part of his game. Changing the locks doesn't stop him—or the people he hires, to be more precise. *He's* too smart to be the one caught. I walk through my home with my jaw clenched in rage. His surrogates do just enough to let me know that someone's been in my house. I don't know how they get in, but there's no sign of forced entry. I hardly speak of it, only to a few close friends. But it changes my life. I feel violated and vulnerable. But mostly I'm angry—at the thought of how much pleasure Paul must get from gutting and rearranging my life.

A few days later, at a coffeehouse, I run into two men who know Paul. One used to be his therapist, and the other was in one of his twelve-step groups. They ask if I'm in touch with Paul. I say no, but explain a little of what's going on. Professional ethics keep him from elaborating, the therapist says, and then adds that he thinks Paul could be dangerous. His bald statement, affirmed by the nod of the other man, doesn't so much surprise as shake me.

I cut even more ties to Paul, no matter how loose. It doesn't matter. Paul keeps fingering my life, making sure I won't forget him.

When my therapist suggests I call the police, I vacillate. When I practice the words I might say, they sound even to me like the delusions of a disturbed woman. I can hear both sides of the dialogue. The power in my apartment is out, but nowhere else in the house or the neighborhood. (It's probably an old wiring problem.) My phone has been disconnected. (Did you forget to pay your bill?) The door to the apartment is open. (You must have forgotten to lock it.) Things have been rearranged. (There's probably a logical explanation.)

It all sounds so unlikely. But, of course, it's happening.

Chapter 35

Though the scope of Paul's harassment amazes me, I don't think of it as stalking. I merely think of it as rage unleashed, circling me and nearly everyone I know. To protect myself, I think, I must turn inward.

It's my nature to live life full strength all the time, which is exhausting. Reading books works as a dimmer switch. Read a book, I say silently to myself when I feel overwhelmed by this situation with Paul. And as I browse the shelves, I remind myself to pick wisely. Not Thomas Hardy. Reading Hardy at a time like this makes one yearn for death. It's easy to imagine almost any Hardy character jumping off a cliff into the English Channel. Lawrence Durrell's not much different, although at least Durrell makes you feel that whatever is happening would be worse if you were an Anglo living in Egypt in the early part of the twentieth century.

Perhaps Anita Brookner, who's meant to be read again and again. Her books are dense with meaning, and each reading leaves something behind—whole sections to be considered, entire sentences to be examined—for the next time. Her solitary characters think much more than is good for them and are far too careful with their hearts. They're proof it's possible to be more isolated than I am. And more timid. Reading Brookner always makes me relieved I'm not timid.

George Eliot, on the other hand, is for security. *Middlemarch* holds special appeal for me, beyond the fact that the story can occupy you for a *very* long time, if need be. I have three copies, one in my bedroom, one in my living room, and one in my car. I prefer to think of Eliot by her real name, Mary Anne Evans, but I understand why she changed it.

These days *Middlemarch* has extra significance for me, because Dorothea chose badly, *very* badly. The husband she chose was nasty, cruel, even more so when he realized she saw through him. But she was lucky: he didn't live long enough to punish her for a lifetime. Anne Brontë's Helen in *Wildfell Hall* lucked out with a death, too. I remember Helen's opening line in a letter toward the end of the book, "He is gone at last."

Yes, thinking about Dorothea and Helen helps me keep hope alive.

Chapter 36

Although we've tried, my relationship with Jen has grown distant over the last several months. We're both sad about it, but it seems inevitable, which is little consolation when I get a phone call from one of her close friends, a former neighbor of mine. "She's got cancer and doesn't have long," she says. "She wants to see you." She doesn't tell me what kind of cancer.

As soon as we hang up, I call Jen and Doug's house, and talk to their oldest daughter.

Mom really wants to see you, she says.

I assure her I want to see Jen, too. Call me whenever she's up to it, I say. I'll come anytime.

I never get the call, but I understand. It's awkward, given the situation with Paul. Less than a week later, I come across Jen's obituary in the paper. I spend the next two days thinking about whether or not to go to the funeral, and decide against it.

The morning of her funeral, I wake up early, wrestling with the question again. Even though I feel rotten about missing it, I can't bring myself to face Paul. I'm not much for naps, but by late morning I give in and lie down on my bed. Just for a few minutes, I tell myself. I fall asleep and wake two hours later with a start: a trail of cool, fresh air brushes the hair on my bare arms. I sit up and look out the window just beyond the edge of the bed. The top branches of the old oak aren't stirring at all. I turn to look at the clock. They should be at the cemetery by now.

Jen's stopped to say good-bye on her way out of town. That's what I think. I'm sure no one in her circle understands my absence, but I know she would, seeing I mourn her privately more purely than I ever could in public.

My friend Ellen is on the board of a new local nonprofit press. She contacts me about a part-time position as the director and publisher. Initially, I decline, but after repeated requests to apply, I agree to meet with Jane, the cofounder and board chair. We meet for lunch at a trattoria a few blocks from my house. Over salads, we take each other's measure. She's younger than I expected, and more serious.

"We'd be so lucky to have you," she says several times over the course of our two-hour meeting. Finally, I agree to take the job.

Shortly after I start, Ellen leaves town to live in Lisbon for a year. It's a sudden decision, and I'm surprised she can afford it. She's talked about it for years, but she always lamented the fact that she didn't have the money to make it happen.

We hold board meetings for the press at my apartment. Before long, Jane begins taking issue with many of my decisions. It confuses me, because my ideas are congruent with those I verbalized in the interview process and during our early days of working together. When the conflict between us comes to a head, the other board members tell Jane they stand by me. "You're making your disagreements personal," says one board member. They take a vote, and all but Jane voice their support of my leadership.

Jane's change of attitude is frustrating, but, in the scheme of things, it feels harmless enough.

Chapter 37

I'm back at the college, teaching part-time and serving as an adviser to the campus newspaper. It doesn't seem as if nearly twenty years have passed since I sat in many of the same classrooms in which I now teach women's studies and writing.

It's a hot afternoon, full sun, no stout clouds for relief, so I'm tempted to stay in my office, which is nice and cool. I was lucky enough to get the big first-floor office at the end of the hall. Lots of windows, lots of shading trees, an enviable spot I landed only because its regular tenant, a novelist, is on sabbatical this year.

But hot or not, I need to stretch between classes, so I put my books and student papers away, tuck some money in a shirt pocket, and lock the office door. I'm not going far, just to the coffeehouse on the corner. I'm almost at the main doorway when the head of the English department, a nun (which you'd guess despite her khaki slacks and white T-shirt), calls my name.

I step inside her office and sit in the carved oak rocking chair across from her desk. She's no fool: rocking settles almost anyone. Like me, she's not capable of a poker face, so I can tell she's nervous by the way she bites her lower lip. The department secretary has had several calls, she tells me, from a man asking about my schedule. Each time he's more insistent. The last time he wanted to sign up for one of my classes, and he doesn't care that we're a few weeks into the term. He's happy to pay for the whole semester. Even audit one of my classes if that's all they'd allow. She assures me they gave him no information and told him he couldn't sign up for one of my classes. She lets out a sigh, clearly relieved to have delivered her message. She's the one who should be sitting in the rocker.

I ask if he gave a name. The department head says no, but tells me the secretary is quite sure the voice is the same each time. The last time she told him not to call again.

"Thanks," I say, aware of the tension in my voice. I don't want to talk about this anymore. I need to be alone. I will myself to appear calm, thank her again, and tell her I'll talk to her later.

I go out the side door of the tan brick building and resist the temptation to stop in the cool comfort of the English garden. Right now, moving is key: I head for the Dew Drop Pond that has stood on this property since long before anyone thought to educate young women here. The campus sits on the second highest hill in the city. The highest point was already taken, by the city's cathedral.

As always, I feel calmer just looking at a body of water. This pond, though not large, has an island in the shape of a drop. Thus the name. I pass the oaks and willows draping the edge of the pond, and cross the footbridge. I turn to look back at the college where I'd felt safe until a few minutes ago. Always lots of people around, students and teachers and staff during the day, security guards at night. I cross back over the bridge, walk down the avenue, and head for the shady side of the street that leads to the coffeehouse.

I move fast, passing the low brick building that's been turned into a retirement home for the nuns who are responsible for the success of this college. I smile and return a wave from the tiny robed nun crouched near the front door pulling weeds. She looks cool and content, sleeves rolled up, long black skirt pooled around her knees. She must be ninety years old, yet her soft face has the look of an innocent.

I pick up speed, trying to keep pace with my brain. It's him. I know it's him. He wouldn't be the one actually calling. But I know he's behind it.

When I get inside the door of the coffeehouse, I take my sunglasses off and step in line. Everything here is cheerful, the yellow walls, the brightly painted tables and chairs, and a blackboard with large white asterisks announcing the day's specials. I stand, half facing the counter, half turned toward the door, and try to shake my feeling of unease.

A man steps in line behind me and places his hand on my right shoulder. "Hello," he says, and calls me by name. He has a face I've never seen before. I'm sure of it. I'm bad at names but good at faces, and in the last two years, I've become even better at registering physical details. Even with sunglasses on, the man strikes me as ordinary—tanned face, no visible scars, short brown hair, medium build. He's wearing faded jeans and a polo shirt. Ordinary, but not familiar.

"Do I know you?" I ask, though I already know the answer. I pull away, but his fingers don't release. Instead, they press harder.

He smiles.

"I don't think I know you," I say. My voice is tight.

"No," he says, his smile widening, "but I know someone who knows *you*." I can't see his eyes, but the rest of his face doesn't look the least bit friendly. His job done, he walks away.

My heart racing, I step out of line and watch him go. He crosses the small parking lot and breaks into a run as he turns down the street leading away from campus. In his wake are children running through a sprinkler on a front lawn and watchful young mothers sitting on the front steps of a mock Tudor. Suddenly this idyllic city neighborhood feels anything but congenial.

Of course he doesn't need to give a name for me to know who he means, who paid him to do this, who told him to be sure to park his car out of sight. Far enough away so there's no license plate to check.

If I'd only known. That's all I can think. But what? What could I have done? What *should* I have done? At what point would all the knowing in the world have stopped this from happening? And what should I do now? I have lots of questions and, unusual for me, hardly any answers.

All I know is that until recently I had only a passing acquaintance with fear. Now it's my constant companion.

Chapter 38

The house on the corner has been for sale most of the summer. On my way out of town one afternoon in September, I notice moving vans outside. When I return ten days later, dusk has already come and gone, and lights are on in the house. As I drive past, I catch the blur of a familiar painting and a man sitting below it. I brake and fight the instinct to back up for a second look. Instead, I quickly drive on, pull into the garage, grab my bags out of the trunk, and run up the stairs to my apartment. I lock the door behind me and leave the lights off.

I sit on the floor in the dark. I go over the scene in my head. There may be plenty of men who resemble Paul physically, but it's far too much of a coincidence that such a man also has a painting that looks exactly like the one that hung in our bedroom, the one he commissioned. I try to convince myself that my conclusion is wrong, that he wouldn't be so obvious.

Shortly after Jen died, I'd heard that Paul married a woman he'd met just before I'd moved out of his house. Under cover of a new marriage, with Jen gone, it appears he can no longer resist the temptation to make watching me as easy as looking out his window.

Until now, I've seen this as a sick game, meant more to annoy than to frighten me. This latest move changes the face of things: Paul, who lives his dark side in such secrecy, is acting out his obsession overtly. The only thing that makes sense to me is this: he knows that more than anything I need to feel safe in my home. It goes back to all the nights when my father, drunk, would throw us out of the house. Sometimes it was me. Sometimes it was one of my siblings. You never knew when your number was up. Not because of anything you'd done, just because you happened to be on his mind and in his line of rage. And if it wasn't your turn, but you spoke up for the target, you'd get thrown out, too.

Suddenly it makes perfect, if crazy, sense: I frustrated Paul's attempts to control me while we were together. I resisted his attempts to bring me back after I left. He hasn't been able to get me to react to all the things going wrong in my life now. Moving to my neighborhood—the one he said he'd never live in—is the only thing he can think of to penetrate my shell.

I get up from the floor, flatten myself against the wall, and approach the windows. I pull the curtains closed, one by one, and recheck the doors to make sure they're locked.

The wise thing to do, of course, is to leave as soon as possible. But I refuse to be rushed. Every time I move, it means disrupting my work life, too, and I'm in the middle of several major writing assignments. At least I'm not teaching or advising at the college now. After the prying calls and the incident at the coffeehouse, I told the department chair and dean of students I thought it best for me—and for the students—if I avoid a predictable schedule for the time being.

I need time to consider my options. I don't know what to do or where I'll go. All I know for sure is that I don't intend simply to move again. Paul could easily follow me to another neighborhood. Before I go to bed, I look in the phone book for a rental place so I can store my things. Then I take my duffel out of the closet and throw in underwear, socks, pajamas, and my jewelry. It's a start.

The next morning, I give the landlady my sixty-day notice, explaining why I'm leaving. She tells me she's putting the house up for sale. Her plan is to tell the neighbors first. I tell her Paul will ask to look at the house. I'm sure of it. "I'm afraid of him," I say. She promises to tell me if Paul requests a tour.

Back upstairs, I mark boxes with the room their contents will inhabit another day: bedroom, kitchen, office, bathroom. I console myself with the fact that this time I'll be packing nothing he's touched. I've already thrown out, given away, or burned everything he gave me. One night, weeks after I left him, I lit a lush summer bonfire in the middle of the woods and raised my eyes to watch the ashes of my life with him float away.

Or so I thought.

Chapter 39

When I tell my therapist about Paul's moving to the neighborhood, once again she advises me to go to the police. He's moved from psychological to physical stalking, she adds. I know it's sound advice, but I don't call. What would I say? My ex-boyfriend just moved in across the street from me. I wonder if you could do something to make him leave. Somehow I can't picture them rushing right over. Even to me, the request sounds absurd.

So instead, I continue packing and focus on my next move. Every spare minute, I organize boxes. Every time I look at an apartment in a new neighborhood, I imagine myself moving in only to have Paul follow me. Meanwhile, as I try to figure things out, I keep to my usual morning routine: coffee, bath, work. At the same time, I change my outside habits: no more stops at the coffeehouse in the next block, no more movies within miles of my house, no more groceries from nearby stores. Whenever I leave the apartment, I always head away from his house, even if it takes me out of my way.

I call a special board meeting of the press and tell them I'll be moving, perhaps leaving town for a while. I briefly explain the situation, and when one board member asks Paul's name, I give it. Within an hour after they leave, Jane calls. Paul's wife is one of her closest friends, she says. I recognize the name as the woman from the women's clinic with whom he had wanted me to arrange lunch. He got her to quit her job, and he's putting her through graduate school, Jane says. Her friends are worried about her. She's changed, she explains. And then comes the kicker: Paul's been asking questions about me for the last several months. He told her that we'd dated casually years ago, that I was interested in him, but that he wasn't interested in me. He says our parting was acrimonious. Well, at least he got the last part right. I suspect she's not telling me everything, but frankly this revelation throws me. I can't get off the phone fast enough.

Two days later she calls again. She tells me that when she told Paul about our disagreements, he offered to make a large donation to the press so she could control it, and fire me. She assures me she didn't accept his offer. "He uses his money to gain control of people. I can see that now. I'm even worried about the influence he has over my husband. They've become friends."

It doesn't end there. Paul's wife, she continues, is increasingly worried about his interest in child pornography on the Internet. She starts to give details, and I stop her. It's as if she's looking to me for help. I don't even know how to help myself. But mostly, I want to get her back on track.

"Didn't it occur to you that he's just trying to make my life miserable because I left him, *and*," I add, "does it occur to you that you've been helping him?"

Now it's Jane who brings the call to a quick close, but within a few days, she calls again. "I realize now that he used to call me right after I got home from our meetings," she explains. He wanted to know about the layout of my apartment, where the bedrooms are, the doors, that kind of thing.

I can't believe she didn't tell me this right away. I tell her so.

She rushes on: "Moving across the street from you was *his* idea. She didn't want to."

Bully for her, I think. She went along with it. All I can think about are dead phone lines, interrupted power, unlocked doors. Jane doesn't apologize. Mostly she just excuses herself: "I didn't think there was anything wrong with me telling him anything he asked," she says.

Again, I express dismay.

"Now I can see I should have been a little suspicious." Her voice has taken on a whiny quality, as if this is happening to *her*, as if realizing she was used somehow justifies her behavior.

I suddenly see my involvement in the press as a setup. I'm more than qualified for the job, but now I suspect that I was hired so Paul could keep me in his sights and manipulate my life in yet another way. I'm furious, but I keep my anger close and controlled. I don't trust Jane, and I hide my feelings from her.

As I hang up the phone, I realize that moving away from here will require even more careful planning than I thought.

College, graduate school, and my months with Paul aside, I've slept in the same bed since I was fifteen. The walnut frame with inserts of burled wood, in the style of Louis XV, is the one my mother slept in and her mother, too. As soon as I hang up after Jane's last call, I take it apart and piece by piece carry it to the spare room on the second floor. The room faces the backyard instead of the street. Besides, it's closer to the outside door, which feels safer now that I know Paul is familiar with the house's layout. I'll sleep here during my remaining nights in this house.

Each morning when I wake, before I even get out of bed, I focus on the most important work of my life these days: how to get away from Paul for good.

Chapter 40

Some days, my sessions with my therapist feel like my only link to sanity. It's how getting a transfusion must feel: depleted one minute, filled up the next. She knows what I need: patience and silence first, so I can verbalize what I know; and then infusions of sound psychology—observations followed by suggestions.

I tell her I can see that many of the clues about Paul were there from the start—now that I know what to look for, of course. Sure, I thought he was a bit unconventional, but frankly, I am, too, and I find a certain degree of eccentricity appealing. Of course, I now see how much of his is simply perversion masquerading as charm. And, I tell her, now I can see there were significant flags that should have warned me.

Yes, she agrees, but growing up as you did, with unpredictable swings between kindness and cruelty, you thought that was normal. You thought that's how all love is.

Still, it's difficult not to feel stupid at the simplicity of it all, I tell her. My father's hateful attacks primed me to put up with anything if a man professed to love me. My mother's loyalty primed me to put up with anything, too, if it was at the hands of someone

I loved. But, I add, it doesn't *feel* so simple when you're in the middle of it.

Bottom line: my whole life led me to Paul. In the middle of a life that is all about hiding from him, I see that I had to pass through him, or someone very like him, to get to the other side of dark. I spent my adult life avoiding what I was raised to do—marry an alcoholic or someone as emotionally damaged as an active addict. Oh, like other daughters and granddaughters of alcoholics, it's never a stated goal, but it's a destination sure enough.

I was trained to forgive anything in the name of love, the name-calling, the blows, the hateful looks. But Paul, unlike my father, was irredeemably perverse and manipulative. Paul never actually hit me. He simply demonstrated how you go about murdering a woman's heart. I left before I was destroyed completely, but I can empathize with women who know how dangerous their men are and are more terrified to leave than to stay.

I'm ashamed that, while I worked so hard to avoid alcoholics, instead, I picked a sexual deviant and traded drunken rages for cruelty of another sort.

Should I worry, I ask her, that I dream about one of my sisters killing Paul? I don't kill him myself, I clarify, as if that might absolve me. But, I add, I do love the dream, and I have it often.

Of course not, she says. It's natural, in fact. You're holding your anger at bay in order to get through this, she continues, and it has to come out somewhere.

It's not just my therapist who helps me be kinder to myself than I would be on my own. My friend Phil, who'd met Paul, reminds me: "How could you have known? He *looked* good. He *sounded* good." And just as I'm about to protest, he adds, "He *did* most of the right things." He interrupts my next objection with a laugh. "He'd charm the laces off an old boot!"

* * *

Just as I'm getting ready to leave for a client meeting one day, I stop, one tan leather glove on, the other on the hall table. I shouldn't come back here tonight. The thought roots me to the hardwood floor. Trusting the instinct, I phone my client and explain I'll be fifteen minutes late.

To my already half-packed duffel I add toiletries and a change of clothes. I toss a few work files into the shoulder bag that holds the papers I need for the meeting. As I leave, I carefully lock the door behind me. I turn the knob to make sure it's secure.

After my meeting, I go to a nearby bank and withdraw several hundred dollars in cash. I check into a hotel several miles from my house and register under a different name. I pay in advance since I don't want to use my credit card. Standing at the check-in desk, with a pile of city newspapers on one end of the counter and a glass bowl of apples on the other, I realize that, in the midst of this ordinary-looking hotel, I'm terrified.

In the room, when I check my messages at home, I have a call from my cousin Maggie. When I call her back, her husband answers. I explain where I am.

"Come and stay here," Greg says.

"I don't want to put anyone else in danger," I say.

He dismisses that and says, "Look. You'll be safe here. Come over now."

I want to be alone tonight, but I'm tempted by the safe haven he offers. "I'll come tomorrow," I say.

Before I retreat to their guest room the next day, I stop to buy my first cell phone, not missing the irony: Paul wanted to buy me one so he could reach me anytime he wanted. I buy one to get away from him.

A few days later, I go home to pack a few more things and pick up my computer. When I get to the door, it's open and unlocked. There's only one other sign that someone's been inside—a book that I know I'd left on my desk is on the stairs, halfway up—just enough to let me know someone's been here. The thought of Paul across the street, fully aware of the layout, makes this seem even more ominous than before. I hurriedly pack the work I can't replace, the Brontë notes, and the rare books I've accumulated over the years. I'm prepared to walk away from everything else.

When I get back to my cousin's house, I finally call the police, even though admitting that I need their help makes me feel weak. I'm only now beginning to see the downside of my hyper-self-sufficiency. The very thing I've prided myself in—not asking for help, taking care of everything myself—may have only escalated his behavior. I hope it isn't too late for help.

Chapter 41

I live in the third precinct. So does Paul. When I dial the number, it occurs to me that the one benefit of having him across the street is that the same police will deal with us both. Hoping to increase my chances of being taken seriously, I ask to speak to a woman detective. I've heard about police who question cases like mine. What'd you do to make him so mad? He must really love you if he can't leave you alone. Why don't you go back to him? Are you sure you're not just imagining all of this? It's bad enough when family and friends say such things, but it would be devastating to hear it from the police, too.

As soon as I start telling the sergeant my story, she interrupts me. "I can take the report," she says kindly, "but you'll get more help if you call the Sex Crimes Unit downtown. Precinct time is pretty much spent with perps beating up women and breaking down their doors."

The woman detective in the Sex Crimes Unit I reach asks if I can come in so she can take my report in person. She adds, "Bring a photo of your ex." I tell her I've already thrown away or burned all my photographs of Paul, but before I leave I call my sister Liz and ask her to mail a picture of him from the last Christmas he spent with us.

Soon I'm in a police conference room facing three detectives, who all look half ready to believe me and half ready to discover I'm a nut. I tell them everything about my time with Paul and all the things that have happened since I left him. It takes a couple of hours, during which time one of them never stops taking notes.

When I get to the kidnapping attempt of my sister's children, they ask me if I think he was behind it. I've considered it by this point, of course, but I don't know what to believe.

I shouldn't waste any more time or energy wondering, one of the detectives tells me. They don't believe in the kind of coincidence necessary to think that the man who tried to kidnap my niece and nephew just *happened* to be someone who used to work for Paul.

I *did* think it an amazing coincidence, but it didn't occur to me at the time to consider that Paul might be involved. I suppose it's natural *not* to suspect that someone you love is capable of kidnapping children. We all bought Paul's theory, which handily kept us looking at him as part of the victim group rather than as a possible perpetrator. Paul could have hired someone to kidnap the children, one detective muses, then hold them for ransom and offer to come up with the money himself.

That would be one way for him to have ingratiated himself with our family, I think. It must not have occurred to Paul that we wouldn't accept his money anyway. He wasn't part of the family, and we had resources of our own.

One of the detectives asks if I believe Paul could have been behind his father's murder. He's the obvious suspect, they say. He was the only heir, and it sounds like he had a terrible relationship with his father. I tell them about the man who's been convicted.

He wouldn't be the first one willing to go to prison after being paid to murder someone, says one detective. You'd be amazed at what people will do for money. And this guy sounds like he has lots of it.

But he got the death penalty, I point out.

She explains how many people never actually get executed: appeals can last a lifetime.

I tell them about the party Paul hosted after the trial, at the country club.

Sounds like he charmed them all, says one. If the police in New Mexico didn't suspect him from the start, they weren't doing their job.

The detectives' response forces me out of my denial about how dangerous Paul could be. If he's capable of arranging the kidnapping of my niece and nephew (which I no longer doubt), the incident in London (which I hadn't thought to connect to Paul), and the murder of his father (which I now consider possible), I'm as expendable as a person can be, especially now that I've gone to the police.

Before I even finish telling my story, one of the detectives asks what took me so long to come to them. This is *not* the response I expected. I haven't been beaten up, raped, or shot, so I've assumed this would seem minor to them.

One of the first things they check out is if I'm still in contact with him.

Not at all, I say. "Not for more than two years."

They're glad to hear it. That's one of the basic mistakes, they say, when you're being stalked. A stalker wants his victim to react, to engage. It keeps the cycle of connection going.

Then one of the detectives, the one who'd introduced himself as having thirty years' experience dealing with stalking crimes, says, "His goal is to drive you crazy."

"To make people *think* I'm crazy," I say. I want to make sure I understand exactly what he means. I focus on his weathered face. He looks like he knows what he's talking about.

"No," he says. "That, too, but mainly to drive you crazy." For clarification, he adds: "To literally make you crazy and have you labeled crazy. Do you understand what I mean?" He stops, but only long enough to register my stunned nod. It's what these men want, he continues. They want their victims crazy. If you're institutionalized, they know where you are, and they can watch you more easily. Do you know how easy it is to mess with someone who's officially considered crazy? he asks. Obviously responding to the not-me look on my face, he adds, "I've seen it happen to women just as smart and well educated and capable as you clearly are."

As if from a long way off, I hear one of the other detectives chime in. Here's how it looks to them, she says: I've lived a relatively harassment-free life; I meet this guy, and suddenly all sorts of things happen to me, way too many to write off as coincidence.

I'm beginning to feel fuzzy, as if my brain is shorting out, or running low on power. I take a deep breath and straighten in my chair. I need to focus on what they're saying. Now is no time to be overwhelmed by feelings.

This conversation changes my life forever. It's a scene out of a movie, a bad movie. But it's happening to *me*, and it's real, and it's forcing me to see the future as I've never imagined it. Someone is out there who hates—or *fears*—me so much that he's willing to dedicate an enormous amount of time, money, and energy to drive me literally out-of-my-mind crazy. Long story short, he'd like to turn me into another version of myself—a form unable to work, to trust, to live freely, to feel safe ever again.

The detectives assigned to my case tell me they'll start looking into it immediately.

* * *

Two days later the lead detective calls and asks me to come downtown again. "We've started checking into your story," she begins, "and everything you told us lines up. Even some of his friends verify what you've said. We hardly ever see that," she says, adding that the details I've given are amazingly accurate.

I have a very good memory, I tell her.

She laughs as she tells me that's what my sister Liz said when she called to check out the kidnapping attempt.

"Some people find it annoying," I add.

She laughs again. "Yeah, she said it's a pain when she's arguing with you about something that happened before. But at times like this," she continues, "a good memory is really helpful." She goes on to say that they've advised Liz to talk to the police in their small town and to all the children's teachers as well.

As often happens, good news is followed by bad. The police are concerned that it's been more than two years since I left Paul and had any contact with him. Typically, harassment is at its worst right after a woman leaves. Evidently, the fact that Paul's kept his rage alive this long isn't a good sign. And because he has others harass and threaten me, the burden of proof is enormous, perhaps impossible. Despite that, she assures me they intend to do what they can.

And so it starts. Now *he's* the target of scrutiny. They run background checks and call people who can corroborate the incidents I've described. They park in front of his house and watch him. "We want him to know how it feels to be watched," says the lead detective, "and we want his wife to know we're watching him." To warn her about her husband, she explains.

Chapter 42

Best not to stay with your immediate family, the police advise. Those are the first places he'd watch. So it's good that I'm staying with Maggie and Greg. They've turned their guest room over to me and hide my car in their garage at night. But more than that, they make me part of their lives and remind me I'm not alone in this.

After I move all my things out of my apartment, the police plan to bring Paul in for questioning. They want me to be as far away as possible because they think he'll be at his most dangerous after that. I can stay with friends out of town, I tell them, while I finish the assignments I've got due before the end of the year. I live on what I make. Now's no time to be irresponsible with my clients. Then I can visit friends in Europe.

I consider my options. He'll guess England first. So I think of Ellen, who's still in Lisbon. Not long after I've gone to the police, she happens to call. I tell her what's going on. I can stay at her apartment there, she suggests. She's coming home for a few weeks over the Christmas holidays. And then when I return home, why don't I rent her house back here, she continues. She'd made this suggestion before, when she decided to make the move to Portugal, but I wasn't interested. It's large, three stories, with lots of outside doors. As for Lisbon, I don't know the city; the idea of going somewhere new isn't comforting. Ellen pushes hard, and is clearly annoyed by my resistance.

I relay our conversation to the police. I also tell them that before Ellen left, she looked at a house that was for sale directly across the street from my apartment. I knew she couldn't afford it, so it surprised me. When I asked her about it, she said she knew someone who might be willing to help. Until now, I hadn't thought of Paul. As far as I knew, they hadn't ever met. Of course, now I wonder. They suggest that Paul could have paid her all along, to urge me to get a job at a place he thinks he could control and to live in places where he could have easy access.

When I call Ellen back as I'd promised, I tell her of the police's suspicions. "But," I lie, "of course *I* don't think that."

"I've got to go. I'll call you later," she says, and abruptly hangs up. I never hear from her again. She doesn't call back, and she doesn't respond to the messages I leave over the next couple of days.

I start to inquire about a place in Ireland, near Cork, but decide instead to go to England after all. I'll fly to London, then stay with my friend Polly. She's sold the B and B and moved to a cottage in a small village in the North.

Paul knows how I think. He knows England would be the first place I'd think to escape to. Then he'd discard it, suspecting I wouldn't pick such an obvious choice. I count on him to think that way.

Chapter 43

Until now, I've never been much for instructions, reading or following them. But I have entered a script of someone else's creation and have no idea what my next lines should be. So when the police tell me what to do to increase my chances of staying safe, I follow their directions down to the last detail. Tell your family, friends, and clients about what's going on. To increase the net of safety around you, they say. But tell as few people as possible where you stay and where you go. Never drive the same route to any destination. Use only secure or public phone lines. Keep a second calendar at home with your up-to-date schedule so that if you disappear we'll know where to start looking for you. Sell your car in case it's got a tracker on it. Set up private bank accounts with extra security. See a therapist to deal with the trauma.

The minute I leave their office, I drive to the car dealership and arrange to sell my car and rent a new one each week until I leave for England. Tracer or not, I learn not to second-guess such decisions. I'm too busy half looking back to see if I'm being followed and half looking forward to decide where to go next.

The one that kicks me in the gut is the calendar. It's easy to visualize how he'd make me disappear. I picture the dirty, secluded building that he rented shortly before I left him. He showed it to me once, when he went to have the locks changed. It's in an industrial neighborhood on the edge of downtown. When I asked Paul what he was going to use it for, he said, "Nothing in particular. I just might need it some day."

The building is on a street where screams might never be heard or, if heard, might not be attended to. I'd be in one room. He'd be in another, probably far away, watching me electronically. It would give him enormous pleasure.

The detectives present three basic options: settle somewhere with lots of security, settle somewhere with lots of security under a false name, or keep moving around.

When they talk about staying in one place, I ask them, "So I can be an easy target, you mean?"

"Well . . ." the detective begins until I interrupt him.

"And you're going to provide protection twenty-four/seven?" I know, of course, what the answer is. There are hardly enough cops for the streets, never mind private patrols.

They can't afford to do that, he says, but they'll watch me as closely as they can.

I know that, I respond, and I don't expect it to be any different. "But since I can't afford to hire round-the-clock protection," I say, "I guess I won't be doing that." Set myself up as an easy target and hope they'd get there in time? I don't think so. I've done my research. I know how many of these stories end.

The math is depressing. More than a million women are stalked each year in this country, and about 87 percent of the time we're stalked by men. Usually it's someone we know—77 percent of the time to be exact—and 59 percent of us are stalked by an intimate partner. Only 55 percent of stalking victims ever report the crime to the police, and only 13 percent of stalkers who are charged are actually prosecuted. Of those, just over half are convicted.

The more research I do, the more discouraged I feel. The average time a woman is stalked by an intimate partner is just over two years. Since Paul seems to be just heating up, I know I'll be on the bad end of the statistics spectrum. And that doesn't even take into account stalkers like mine who have an added benefit: limitless resources. With stalkers who can afford to hire others to harass and terrify their prey, the hunt can last longer, forever even—the prey's forever anyway.

The more the police tell me what they think of Paul, the more convinced I am I'll be safest if I stay on the move, at least for the foreseeable future.

It's clear I need a world that's spare, with few possessions, but I also need a world full of professionals. I'm tempted to have a tracking device implanted under my skin. I've read about them. I begin to wear a pouch with my passport, credit cards, and cash all the time. I don't even take a walk without wearing it. When I sleep, it's within easy reach. Being ready to run makes me feel less vulnerable.

Chapter 44

I've given most of my things to a shelter for battered women, so my apartment is fairly empty. But I still need to meet the movers and store what's left—boxes and the furniture I'm keeping, mostly family antiques. A few friends meet me at the apartment, and we finish packing.

The early morning rain turns to sleet, then halfhearted snow that covers everything in slush. As I leave the apartment to show the movers the way to the storage unit, I drive past Paul's house for the first time since the night I saw him in the window six weeks ago. When I glance inside now, I see a thin woman with a lonely face. It's no contest: I'd rather be outside and this afraid than inside not knowing how afraid to be.

As the truck is unloaded and our cars are emptied of boxes, my sister-in-law Sally and I stand calf-deep in icy water. Getting soaked and wondering if this unit will be locked when I return are the least of my worries.

I call the police to tell them I'm on my way out of town to say good-bye to my mother, and Liz and the children. Now that I've done as much as I can with the police—at least for now—and I have a break in my deadlines, I can leave town for a few days.

They're meeting me at a motel a couple hours' drive for each of us.

My mother works at not crying. She does it well. I know the look: her eyes cloud, making the blue less intense. While Liz watches her children swim, we talk. "I could kill him myself," Mom says. "If your father were alive, he'd . . ." She lets the thought hang. When my mother's *really* angry at Paul, that's what she says, never completing the sentence. We don't actually know *what* my father might have done, nothing illegal, but *something*—he would have done *something*. We're sure of that.

Missing my father and feeling powerless against this mad external force that floods our lives, my mother and I make jokes. If we were a certain kind of family, we laugh, we'd be figuring out which new construction site would be best to bury a body in. But even if we were capable of such a thing, our guilt would drive us to turn ourselves in. We also know we're not the kind of people who would fare well in prison. We're not sure who is. We just know we're not. We joke about this, too.

We both know our words are meant only to make us feel more in control than we are. "You're sure you have enough money?" It's a question she asks each time we talk. If my father were alive, he'd be asking, too. I assure her I do.

By this time, she's recognized her part in the damage her children suffered in our alcoholic home. She's seen how she harmed us all each time she told us to bury the hurt and pretend we hadn't been sworn at, punched, or simply battered with insults. She's seen what she asked of me as her eldest daughter each time she made me act as another parent instead of holding Dad accountable. Her words of regret are backed by days such as these: she knows it's never too late for a mother to be present for a child, no matter how old either is.

"I promise to tell you if I need anything," I say, and then rush on with what is painful for both of us. "You know I won't be able to tell you where I am, but I'll call once a week to tell you I'm all right." The police suspect her phone could be tapped, so we agree on a code for my calls, one impossible for anyone else to interpret, except perhaps another English major as well versed as she is.

When we part, she wraps her arms around me and holds me to her, the way mothers do when they realize they can't protect their children from harm. It reminds me I'm still someone's child. I'm not as experienced in pain as she is. I'm the one who cries.

Chapter 45

The police want me to stay out of town as much as possible until I leave for England, so I've made plans to stay with friends who have a B and B an hour west of the city. On my way there, I get a call from my cousin, Dan, and his wife, Christine. Since I was in my teens, Dan, who's eight years older, has been more like a brother than most of my own. And Christine has been my friend for twenty years. They've called to see if they can do anything to help. I tell them I'm leaving town, but since I'm on my cell phone, as are they, I'm not specific.

"Do you need a full-time bodyguard?" Christine asks.

"Some days I'd love one," I respond, "but I can't afford it."

"That's not what we asked," Dan says. "Do you *need* one?"

"Because," Christine chimes in, "if you do, we'll see you get one. Don't worry about the money."

And here's the thing: they mean it. If that's what I need, that's what I'll have. Somehow, they'd make sure of it. But this could go on for years, and I must learn how to get by. So, through tears of gratitude, I thank them and decline. But I know, from this moment on, that I have a wide circle of protection. All I have to do is ask for help—or simply accept what's offered.

For weeks, the friends who own the B and B provide me with a guest room, a desk, and hours of conversation and laughter. Equally important, they share their outrage: they're more angry about the situation than I can afford to be. If I give in to my anger—or fear, for that matter—I'll lose credibility, and business. But more dangerous than that: I'll lose control. I must cap my feelings until I don't need such a clear head. In the meantime, it's helpful for *someone* to feel the rage.

Once I get settled in, I call the police. The lead detective tells me that after several hours of parking squad cars outside Paul's house, they went back to ask him to come in for questioning. They planned their visit for a time they knew both he and his wife would be home, the sergeant explains. If she didn't see them herself, he probably wouldn't tell her they'd questioned him.

When Paul came to the station, he brought an attorney with him. He's so arrogant, she adds, he didn't even bring a criminal lawyer. The guy was a corporate attorney who didn't know a thing about the law involved.

I find out Paul did a classic stalker thing: tried to turn the story on me. She's stalking me, he told them. His face, apparently, was red, bursting with rage, and he was extremely nervous. He kept drinking from a bottle of water he brought with him. I never even loved her, he protested. The police pointed out to him that they already had proof that we'd lived together for close to two years.

Besides, the lead detective said to him, your relatives told us living together was *your* idea. And why would you move across the street from someone who's stalking you? Paul had finally landed in a place where his lies were not believed. I don't even think his attorney believed him, the sergeant tells me.

At the end of the interview, they warned Paul: if anything happens to her—or to anyone in her family—anywhere in the world, you'll be the only suspect. Once again, I'm overwhelmed by the support I'm receiving in the place where I frankly least expected it.

Chapter 46

Habits are easy to break when you're scared enough. Usually when I travel, I book directly through the airline and charge the ticket to my credit card. If you want to be hard to find, there's another way—at least there was before September 11. You go to your bank and take out enough cash to pay for your airfare and several nights in a hotel. Then you go to the nearest airline office and buy a ticket under a name close enough to get you through with your own passport, close enough to call it an error if questioned. You use your middle name as your first name, or transpose your first and middle initials. That's what I'd done when I flew to Maine in October.

I can't take any chances this time, so tucked inside my passport is a letter from the police explaining I'm the victim of an ongoing crime and need to travel under an assumed name. The detectives have already talked to the head of the airline's security department. All I have to do is go to the downtown airline ticket office and pay for my ticket with cash, using the name we've agreed upon. It's to be a name known only to me, to the detectives, and to the head of airline security. I have the cell phone number of the head of security in case there are questions the day I fly to London.

After picking up the letter, I walk the few blocks to the ticket office. While I wait my turn, I scan the blue molded plastic chairs every few minutes. It's what I always do now. "I'm afraid I'm becoming paranoid," I had just said to one of the detectives.

Paranoia, he said, is a healthy response to what's happening to you. It can keep you safe.

When I go up to the counter, I quietly explain the situation. I whisper the words "London Gatwick" when I hand over the letter from the police and ask the agent, under my breath, not to repeat the destination.

As she waits for the reservation to appear on the computer screen, I turn around for another quick scan. Sitting in a chair several yards away is the woman from whom I rented my last apartment. She must have just come in. The detective told me that when they'd questioned her, she admitted talking to Paul about me. She told him I was afraid of him. They don't know what else occurred, but it seems they met right after I asked her not to let him in my apartment.

She's sitting close enough to overhear our conversation now. I turn back to the counter and internally review what we've said so far. Neither of us has mentioned anything specific out loud. I lean in close to the agent and say quietly, "A woman my stalker knows is sitting behind me. I need to make sure she doesn't know where or when I'm going." I stay calm, but I'm sure I look frantic. "Please help me keep this secret," I say.

"Don't worry," the agent says quietly. "I can do this." She finishes the paperwork and slips a piece of paper across the counter. It's what I owe for the ticket. I put down the exact amount, and, as she hands me the ticket, she says, in a loud but normal voice, "It's been really warm there, so be careful. You look like you're not used to tropical sun." She beams as she speaks, as if she's sending me off on a vacation I've won as a prize.

I've been holding my breath, but I beam back as I let the air rush out of my lungs. "Thanks for the tip," I say to her in as natural a tone as I can manage. Before I turn away from her, I mouth more gratitude: *Thank you. Thank you.* My lips form soundless syllables. I place my hand over my heart.

I walk away, keeping my face turned from my former landlady. Before I'm out of the room, the ticket agent says, "Say, don't forget to bring plenty of sunscreen." As I turn my face in her direction, I can barely see her through the tears that blur my vision. But I can see her clearly enough to know that she reads my tears as the greatest compliment of all.

"I won't," I throw back in her direction. "Thank you."

I walk down the street. Now I'm really on the run, I think. I'm at once terrified of what's ahead and overwhelmed by the kindness of a stranger. I need to believe I'll know such kindness again, because the police are right: Paul has enough money to track me if he wishes—anywhere in the world, for as long as he wants.

Chapter 47

As I wait to board the plane to London, I repeat my false name over and over in my head. You have to choose one difficult to guess but familiar enough to answer to, even when asleep or relaxed (a relative state when you're being stalked). If you pick something with no connection to your past, you tend not to respond.

As soon as I'm on the DC-10, I toss my duffel on my assigned seat and scan the rows to see if I recognize anyone. I'm prepared to get off if I see someone who knows Paul. I don't recognize anyone, so I stash my duffel in the overhead bin and my backpack under the seat ahead of me. As I step over my two seatmates, I present my best standoffish face, the one that says, I'm not interested in you, so don't even *think* of being interested in me. If need be, I'll pretend I don't speak English. I just want to be left alone as I wait for London, with its libraries, bookstores, museums, and galleries. And tea. Lots of tea. There's something about tea that makes solitude perfectly palatable.

I want to take London back for myself, with new memories that are mine alone, free of Paul. I doze on and off for the next seven hours.

* * *

The crowds are thin in the middle of London on Boxing Day, and nearly everyone, Christian or not, looks to be recovering from the surfeit of Christmas. People are huddled in wool coats as if it's terribly cold. Fresh from a below-zero day, it feels mild to me. I unbutton my jacket and order a cappuccino at a kiosk in Victoria Station and head for the Thomas Cook window. I find a good rate at a small hotel a block off Oxford Street. I've never stayed here before, not with Paul, not alone, not with anyone. London is a city of seemingly endless neighborhoods. Samuel Johnson got it right when he said, "He who tires of London, tires of life." I have much still to discover about my favorite city.

Using my fake name, I check into the hotel, paying cash for a few nights in advance. They want to keep my passport in their safe. I'm not about to let it out of my sight, so I show the hotel manager the letter from the police. He lets me keep it. From now on, I'll be ready to leave anywhere, anytime, with no notice.

My first destination the next morning is the British Library, where I expect to feel safe, sane, and productive all in one stop. With several deadlines behind me, I'm looking forward to days of research dedicated to the Brontës. I set off down Wigmore Street, passing beautifully restored Georgian and Victorian buildings. I cross over to Goodge Street and angle my way toward Bloomsbury.

"Sorry," says a woman in a maroon wool cape. I'm the one who should apologize. She's bumped into me only because I'd stopped suddenly to look up at a particularly interesting house.

"No," I say. "My fault. Sorry." And with that innocent interaction, I begin to shed the baggage I've carried for weeks, the feeling of being watched.

I round the corner to Montague Place and head toward the back entrance of the British Museum, which houses the British Library. On my way to the Reading Room, I make my way through the Egyptian collection, pass the Rosetta Stone and a bronze of the Hindu god Shiva. Finally, I'm in the hall outside the room that no one can enter without a pass, which isn't easy to come by. Clever as he is, Paul's surrogates would have a difficult time securing the required letters of introduction and proof of scholarship, at least in time to find me here. I smile at the man behind the desk and present my reader's pass. Almost there. I'm almost there.

"Oh, you're from the States," he says. Some Brits still call it that, as if we're not a real country, just a renegade group of states. So, he continues, drawing out the word, I won't have known that everyone needs a pass with a photograph now. He explains that all I need to do is bring back a photograph of myself. That and handing in my old pass will get me a new one.

Such a simple thing, but one that nearly defeats me. I've used every ounce of energy and willpower to get this far. Just when I think I'm steps away from a place where no one can harass me, I'm turned away. My muscles knot themselves back up. "*Now?* I need one *now*?"

Yes, he says. He's sorry, but it's for security. He can't let me in without a pass with a photograph. With an apologetic smile, he adds, "No exceptions."

"Not even once?" I ask. "Just for today?" No matter how hard I try, I can't keep my voice from sliding into panic. "I'll get a new one tomorrow." I focus on his long brown hair (leaning toward the color of bark and in need of a wash) and his crooked tortoise-shell glasses.

"Sorry, luv," he tells me. I can have a photo taken at any passport shop. Or Boots. There's a Boots with a photo booth on Oxford Street, he says brightly. It's just a few minutes away.

I know where he's sending me. I've been to this Boots before.

He's right. It isn't far. But it feels like an *enormous* detour. I offer him a weak smile as I mumble my thanks.

As if programmed, I walk back down the narrow hall and sink onto the wooden bench outside the Reading Room. I close my eyes, lean back, and lay my head against the cool stone wall. After a few minutes, I rise and walk on to the wide portico of the museum's main entrance. I aim myself toward Boots. I want to cry. I want to sleep. Hell, I don't know what I want. Actually, I do: I want to go back to a time before I knew this maniac.

The photos from Boots reveal the fixed eyes of a tracked animal caught in the scope. I may be thousands of miles away from home, but I carry the hunt with me. I'm tempted to go back to the hotel, crawl into bed, and hide—as if I'm not hiding enough already. But stronger than this temptation is the will to put one foot in front of the other and not be defeated. He *will* not win. I *will* not give in. I *will* survive this.

I stuff the photos into my pocket, put on the face of a pleasant person, and return to the British Library. I write out a request for a manuscript and wait at a desk with a green leather writing pad worn thin by hands before me. Perhaps Virginia Woolf or Oscar Wilde, or Karl Marx or Gandhi wrote at this very desk. Next year we'll all be sent to a new high-tech library on Euston Road next to St. Pancras. The move is necessary, but it'll break a chain of hands writing books on these leather pads safely tucked away under the blue and ivory and gold rotunda that makes me feel as if I'm writing to the heavens. It will never be the same.

Chapter 48

Trains alone could draw me to Europe. Perhaps it's the scenes of ordinary life that flicker past: carts of groceries being dragged home, vegetable gardens struggling to thrive, ancient castles falling to ruin. After a few days working in the British Library and watching people who aren't watching me—in bookstores, galleries and cafés—I catch a train north to visit Polly.

A few days after I arrive, on the way to a New Year's Eve party at her brother's house, Polly asks if she should introduce me by my fake name. "Just use my real first name," I say. "It makes me feel safer." And as it turns out, I sit next to a man who does a good deal of business in America, especially in the city where one of Paul's companies is based. It's a remote possibility that he'd know Paul, but I'm glad to be known only by a first name. Just in case.

Long after midnight, we drive back along a country lane worthy of a postcard. It's snowing so hard her small blue Polo makes the only tracks, though we know there's plenty of other traffic.

"My brother's friends think you're on the run because you're a gangster's girlfriend," Polly says.

"They've been watching too much television," I say, and we laugh.

But then she turns serious: "You have to admit, your life *sounds* bad."

Back at her house, a sturdy stone cottage built in the last century for Yorkshire farm workers, we start the New Year in earnest. In the sitting room, she lights a fire and puts a ceramic bowl, scraps of paper, and two pens on the low oak table in front of a comfortable sofa the color of butterscotch. "A fresh start," she says, "for a better year." Sitting on the floor next to each other, with our backs against the sofa, we write down our private wishes for the coming twelve months. We say them aloud to each other and burn the paper.

Safety, work, contentment. Mine seem like modest enough desires.

To take my mind off things, Polly's come up with a tentative itinerary for my visit. You can tell she was a nanny in a former life. Like a mother in charge of a school holiday, she's researched activities she thinks will interest me. We have an outing every day. Our favorite is Saltaire and the Hockney paintings that hang casually in this one-time salt mill, as if they're not the most valuable commodity for miles around. I stand before *The Other Side*, trying to see what Hockney means by these bold shapes and colors. I'm drawn to one side of the painting more than the other—particularly by the trail of red circles—but can't sort out why.

Several days after New Year's, I move to a B and B closer to the York libraries where I plan to do more Brontë research. The first morning I walk miles through a suburban tract to the York University library. I've imagined this university to be as lovely and charming as its medieval walled city, but for the first time in my life a library fails me. It's a dark day, under a seamless sheet of iron-gray clouds, but even a full summer sun wouldn't brighten the dreary architecture. Inside is no better. Institutional tables and chairs, glaring lights, and metal bookshelves would all dim even the brightest mood.

Since I'd simply stay in bed if my spirits were any lower, I search out the city library, which is old, full of stone and wood, with light fixtures meant to please the eye as well as illuminate the words. In newspapers from May 1848, I read about markets, hangings, and concerts during the days of Anne Brontë's last visit to York, en route to Scarborough.

I return to this library nearly every day of January. Early each afternoon, I cross the Ouse River and stop at a small shop whose front window is adorned with juicy roasts. Just one more lunch-time customer, I work my way to the counter and ask for a sand-wich on a freshly baked roll, sometimes turkey, sometimes pork. Most days it's warm enough to sit on a bench in the Museum Gardens, eat my sandwich, and drink from a thermos of tea. One sunny day, I unwrap my striped silk scarf and lift my face to the bright sky. As if suspended from skyhooks, fat white clouds with yellow-gray underbellies drift across the city.

Most days I sit in the part of the gardens near the abbey ruins and think of the tales of Robin Hood fighting the corrupt abbots of St. Mary's. I feel at home in this world of ancient tales and literary research.

Through Polly, I've met other women who share my interests—literature, film, art. In the following few weeks, between my retreats into books, they wrap their arms around my pain and offer moments of normal life: walks in the countryside, tea around fireplaces, conversations across tables. I begin to believe in the possibility of daily life not tainted by my stalker's insanity.

One of my new friends is a writer and a professor of English. One evening, after we'd all met for tea at her home, she disappears into the bedroom and comes out with a slim gray-and-red paperback. "Usually I'm shy about giving someone my novel," she says, as she extends her hand.

"This is your book?" I ask. Patrice had been introduced by her ex-husband's name. I didn't connect her with this novel.

She nods her assent.

"I've not only *read* this book," I say, holding it to my chest, "I've taught it in a women's literature class." Finally, a coincidence that's pleasant. We can hardly believe the way our lives have crossed, not only now, but before, through literature. In such respites, entirely free of Paul, I see that the trick of my life will be making sure I don't miss these moments of joy.

Chapter 49

One day I break my lunchtime pattern and walk into the York Minster itself. In all the years I've visited this city, I haven't warmed to its centerpiece: the largest Gothic cathedral in Europe. I want to appreciate it as so many have before me: Anne Brontë, for one. So far I haven't succeeded. I tell this to a sculptor I've met, and she says: "Go to the Minster's Chapter House and find the faces." So, I do. Evenly spaced around the octagonal, high-ceilinged room, just above eye level, are miniature stone faces. No two look alike. Some are beautiful, beatific even. Others are grotesque, smacking of evil. These faces capture the entire world. I'm a convert.

On my way out, I see a sign for a healing service scheduled to take place two days later. "All Are Welcome," reads the sign. Two days later, at the library, I keep checking my watch. I have many reasons *not* to go to the Minster service. For one, I'm not sure I believe in such healing. Besides, my research is going well. *And*, I remind myself again, I'm not sure I believe in this kind of thing.

Just before twelve-thirty, I override my doubting self and leave the temporary shelter of the library. At the Minster, I slip past the ropes sectioning off the area for the service and drop into a back pew. I *need* such healing, believe in it or not.

I want to be freed of the despair that threatens to sink me, so I fish for comfort in the words of the ministers—some lay, some not—not one wearing the face of condescension or arrogance so often found in such places. After the prayers, we're meant to approach one of the black-robed ministers. Debating whether I'll leave my pew until it's almost too late, I'm the last in line. I kneel before a round-faced man with rough cheeks and well-worn hands. He studies my face and asks me to name the nature of my illness. I tell him I'm being stalked and feel safe only when I'm far from home.

"*Are* you safe?" he asks. "Have you gone to the police?" Only when he's assured I've done all the practical things does he lay his open hands upon my head and tell me to keep hoping for a better day. "I'll pray for you," he says, "and not just today."

Back in the pew, I have a moment of clarity. At a time when it's wise to trust almost no one, the lesson is to trust wisely. Coming from my volatile home, trust had been a matter of all or nothing. In order to stay safe, I'll need to learn to trust slowly, so when I'm wrong, I'll need to make only a minor adjustment. I imagine this might also make me a better judge of character.

When I return to London two weeks later, I retreat to the British Library once again. Punctuating my hours as a scholar are walks—miles every day—around squares, through neighborhoods, it doesn't matter where. They're aimless, except for their common study: the truth of my relationship with Paul, for I am ruthlessly determined to turn my pain into a future.

A few days later, before I leave London, I visit a painting. I've checked ahead of time. It has moved since my last trip; previously at the Serpentine Gallery, it's now at the Tate. The artist gives us a grotesquely featured family: a daughter pulling off her father-general's tall black boots, the father sitting on the edge of the bed, the mother sprawling behind him, and a son watching it all.

I feel as if I've met them, at a reception for a foreign dignitary. In Portugal, perhaps. These people intrigue me because they look both vigorous and dangerous. I remember bright blue and red and yellow, a splash of green, and rays of pink shining through an open window.

When I find this family again, they aren't at all as I remembered. In fact, I've put two Rego paintings together: *The Family* and *The Policeman's Daughter. The Family* has two daughters, with hair pulled back from faces long and sharp. One daughter is pressing against the front of the father, and the mother (with girlish hair tied with a pink bow) is pulling at his jacket from behind. The other daughter, standing in front of an open window, looks on, with hands folded, as if in adoration. There's no son, but in the corner of the room is an icon of Saint George slaying a dragon.

The boot being shined by a girl, the policeman's daughter, is in another painting altogether. This daughter has the same long face and pulled-back hair that makes her not quite a child but not quite a woman.

Only grays and browns and open windows tie the two paintings together. The only thing I got right is the brilliantly rendered grotesqueness. It feels important not to get it wrong again, so I stand for several moments committing the paintings to memory. As I do, they attach themselves to Paul. Perhaps it happened just this way: when I met him, I walked away with more than half the story wrong. And when I was with him, I didn't pay close enough attention.

Only now, visiting reality, do I see what's really in the frame: he plied me with an enormous temptation to pity him, and when he had me, he held me down and gorged on my weakness. The flaw in his plan: he didn't expect me to have the strength to slip out of his bed before he was finished with me.

I wonder what else the camera of my yearning heart got wrong.

Chapter 50

If money were no object, I'd stay in England for a very long time. But I live on what I make, and being on the run is expensive. So I fly home and try to pick up my life again. As arranged before I left for England, I go back downtown a few days after I return to talk to the detectives on my case.

The police ask me again if I think Paul could have arranged his father's murder. I say I don't know, but I also say what I believe: that's he's capable of anything, because he doesn't know right from wrong.

"Chances are you know him better than anyone else at this point," says the detective. "He's not the type who'd like someone knowing so many of his secrets."

I'd been thinking about this while I was in England. No, I think. He wouldn't like that at all. Even Jen recognized this. And to someone like Paul, that makes *me* dangerous.

Just before I leave, they have one more thing to tell me. I can tell by the way their faces shift that they think *I* could go red with rage this time. "That goes for you, too," one says. If anything happens to him, I'll be a suspect. The only difference is, I won't be the *only* suspect. He's the kind of guy who has lots of enemies, she adds.

I offer up a feeble laugh. Fair warning, I think. I tell the sergeant I get the point. I think it surprises her I don't take offense, but it would be disingenuous of me to be indignant. After all, each night I listen to the statewide news, and when they describe an accident (a car wreck, a small plane crash), I hold my breath with the anticipation some reserve for the lottery. Yes! Yes! Maybe this time I'll be lucky. Please God, I pray (a figurative thought, not a spiritual plea), let it be Paul.

I suspect I should feel more ashamed than I do to enjoy such thoughts. Perhaps in time I will.

Chapter 51

I'm determined to shape my life into something recognizable. After all, this is the city I live and work in. I decide to live at a friend's house, at least for a while, but it feels safest to make "home" a movable concept. I buy a new car, a small SUV this time, so I can more easily pack my car full of clothes and books. I move into my friend's house with the tall candlestick my great-grandmother carried on the boat from England, a small Waterford bud vase from my mother, a bracelet from my father, and the travel alarm Liz gave me for Christmas. These few items become my portable home.

It's a modest neighborhood, mostly with story-and-a-half bungalows built after World War II. Some yards, like my friend's, stretch down to a creek that's a tributary of the Mississippi River. At least I have that one tie to my past.

To the research materials I'm accumulating on stalking, I add books by Gavin de Becker and John Douglas. I read *Obsession* by Douglas twice. Aptly named, it clearly defines stalking. "Any sexual predator, before he's been in the business for too long, has his own preferences and learns his own techniques. He knows how to locate and identify—profile, if you will—the victim of preference, the victim of opportunity. He knows how to get inside that victim's head and create the effect he's looking for: the manipulation, the domination, the control of that individual, then the manipulation, domination, and control of the law enforcement personnel trying to neutralize him."

I've read Gavin de Becker's *Gift of Fear* more than once, too, and I'm making my way through J. Reid Meloy's *The Psychology of Stalking.* It all comes together. When I look at a list of a stalker's traits, I can see that Paul fits the profile to a tee. He qualifies for nearly every quality or behavior associated with a stalking personality. He was too intimate right away. He was too persistent too early. He was overly nice. He was too charming. He was too generous. He was controlling. He knew how to punish when people didn't bend to his will.

When I read about the different types of stalkers, I have trouble putting Paul in a category. Not because he doesn't fit the descriptions, but because he fits *all* the descriptions. Each passing year brings me closer to one truth: if I'd been honest with myself (*and* known more), I would have been afraid of Paul from the beginning. His wooing was fast and furious. He was unfaithful, and ultimately he didn't bother to hide it. He used my past against me. He used cruelty to control me. He was so good at lying to himself, he lost track of what was real—if, that is, he ever knew.

What I couldn't have predicted is where leaving him would lead. I couldn't have guessed he'd have an internal to-do list, each item starting with the same verb. *Hire* someone to leave threatening messages on her voice mail. *Hire* someone to break into her house. *Hire* someone to follow her. *Hire* someone to make silent phone calls to her mother. *Hire* someone to kidnap her sister's children.

I never could have imagined such a litany of malice attached to me. Nor could I have foreseen that he'd be so enraged by his failure to destroy me while we were together that he'd dedicate years to making sure I wouldn't feel free of him after I left.

He's the perfect role model for long-time stalkers. He craves control, even while pouring tremendous energy into cultivating a "laid-back" image. He has the money and the resources to stay removed from the personal danger as well as the personal consequences of his obsession. His family money protects him from the ordinary: work, responsibility, reality. In time, he, like chronic liars everywhere, won't be able to recognize the truth of anything. No matter what else I let go of, I try not to let go of this newfound knowledge.

But here's what all the facts and all the psychology can't tell you: what stalking *feels* like. Only when you're in the middle do you know that it's like being trapped in the manic melody of *Boléro*, building and expanding, constantly repeating itself, with no apparent release to interminable crescendos. But unlike a Ravel performance, the madness of stalking can go on forever.

No amount of knowledge can prepare you for that.

Chapter 52

Spring comes to the Midwest in fits of snow and starts of warmth. This year, especially, the winter seems never to end. I give a workshop at a writing center. Someone in the group begins asking me inappropriate personal questions—the name of my bank, of my insurance company. I try to deflect his questions, but he persists. Do I ever use a lawyer? If so, what's his name? Finally, when I challenge him directly, he leaves. An hour later, when I get into my car, it's unlocked. The battery in my two-month-old Honda CR-V is dead. Reporting something like this seems like a waste of everyone's time. There'd be no fingerprints. What could the police do about it? Nothing, I think, so I call AAA instead.

Some days, I notice a car parked across the street from my friend's house. The driver—a different one each time—never gets out. When I drive away, he follows me. I take this as a sign that it's time to move on, so I stuff my car with duffels and boxes. My duffel holds slacks, sweaters, socks, underwear, and pajamas. A large tote is full of books and notebooks. I put them in the back of my small SUV. The backseat is already covered with boxes of files and books. Each time I park it I wonder if my life will be rummaged through when I get back.

I begin moving between family and friends. Some days I can't remember where I'm staying. I walk out of a client meeting, go to my car, turn the key, sit, and focus on where to go next. I often can't recall for a moment or two where I slept the night before and whether that's where I'm meant to return.

Mornings are a little better. I'm rested, and all I have to do is look at my calendar and think about which route to take from my temporary "home" to a familiar location—an office for a client meeting or a library for research.

But the point is this: absolutely nothing is familiar. Everything is capable of disorienting me, and all without warning. Not even the simple act of getting in my car, heading in the right direction, and going somewhere (anywhere) is automatic. I have to think through each block, each stoplight, and each freeway exit.

The energy required to be conscious of every step—not to mention whether I'm being followed—takes an enormous toll. Some days, I can barely make it through without scraping the wall of despair. On these days, I return to wherever my fleeting home is and stand under a long, hot shower. And as water pours over me, I ask myself what I would do if I had a life-threatening disease or condition. I would face the facts: I survived for one more day. I have no guarantees for tomorrow or next week or next year. Nor do any of us. I'd work to feel grateful for every day. And so I do.

That's all anyone can ask really, life in increments of a day at a time. The only way to survive is to remember that at least for the moment, I'm safe. I go to bed. I rest, comforted by the love and strength of a handful of people, and my books. On good nights, I sleep the sleep of the innocent. On bad nights, I run from him all night long, through empty parking ramps, on country roads, down dead-end alleys. Sometimes I'm alone. Sometimes my mother and sister are with me. We're always running, running away from someone. I can't always see who it is.

Chapter 53

Invitations have been trickling in, even from cousins with whom I'm not particularly close. Just before Easter, I accept one from a cousin whose family will be out of town for a two-week spring vacation. This is the eleventh place I've stayed at since I left my apartment six months ago.

I use their home as a transition camp, between seasons, with the living room as a staging area. I set out the large plastic boxes with snap-on covers I've taken out of my storage unit and replace the winter clothes in my duffel with long-sleeved T-shirts, cotton sweaters, lightweight slacks, and jackets. Into the boxes for the storage unit goes winter: wool sweaters and slacks, warm socks and scarves. On the dining room table are receipts, calendar pages, bank records, and phone bills from the previous year. Each pile is topped with a handwritten record of the subtotal, waiting to be added and recorded on the correct line of my tax worksheets.

On Easter, I roast an organic turkey breast and steam fresh asparagus. I open a bottle of Chardonnay and pretend life is normal. Each night, I slice more turkey, cook fresh vegetables, and pour a single glass of wine. I end each dinner with an oatmeal cookie packed with dark chocolate, a specialty from an organic bakery nearby. This mealtime routine provides one piece of consistent action each day.

I'm grateful to be here, but this house is entirely too cheerful for me. Bright floral prints are everywhere—on chairs, couches, curtains, bedspreads—making me feel pressured to be happier than I am. And then there's the empty swimming pool in the backyard, which feels ominous under a moonless sky. Unclad windows on either side of the front door are wide enough to punch a fist through if you want to unlock the door without a key.

The master bedroom feels like the most serene room in the house (not a print in sight), so every night I go upstairs at dusk and lie in bed to read and think. I'm preoccupied with where to go after my cousins' return next week. One night, when I turn out the bedside lamp, I stare into the dark room and try to evoke an image of my next haven. I fall into a restless sleep but within minutes am jarred awake by the jangling of metal. It sounds as if it's coming from the front door.

Keeping the lamp off, I reach for my travel alarm clock and sweep its small built-in flashlight over the top of the bedside table. The phones I usually lay there, my cell phone and the house phone, are missing. The only night it's mattered here, I've forgotten to bring them upstairs. I turn my fear into anger at myself. There's no other phone upstairs. The day I arrived, I checked the house to see where the jacks were.

In the dark, I crawl out of the bedroom and press myself against the wall in the hallway as I move toward the top of the stairs. I'd left a front hall light on, for security, but it will only expose me if I go downstairs; on either side of the door are floor-to-ceiling windows. I can hear the screen door being jostled. Fortunately, I'd locked that door, too. Every night before I come upstairs, I check every door and window.

I ease my way back along the top few steps and hall until I feel the door frame of the master bedroom against my back. I reach the chair where I laid my clothes. I put on jeans, T-shirt, socks, and shoes, and I reach into the duffel for a sweater. I grab my backpack. The car keys are inside, along with the pouch full of credit cards, cash, and passport.

Relieved the bedroom faces the backyard, I pick the window offering the cleanest jump to the roof of the garage, and calculate how to jump so I might sprain but not break an ankle. With luck, I'll do neither. Actually, with luck I won't need to jump. I feel how to raise the screen, then crouch beside the window. I'm ready to jump the second I know I'm not alone in the house. I wait for a successful smash of glass. I remain silent, knowing that in the quiet April night any noise I make will announce my whereabouts to the intruder.

After tiring of trying the front door, he (I'm sure it's a he) goes around the house to have a go at the back door. He also tries several downstairs windows. Finally, he gives up. I didn't hear a car drive up, and I don't hear one leave. I know he's gone only by the absence of sound. I sit under the window in the bedroom for another two hours and keep my ears strained for any sound. I'm good at this. I think of my adolescent nights when I sat on the floor of my bedroom, listening, ready to rush downstairs at a moment's notice.

At first light, I go downstairs to get my cell phone. I consider phoning the police, but I'm giving up hope that anyone—especially a new set of police in yet another suburb—can do anything to help me. Back upstairs, I lie on top of the covers and try to rest. At seven, I throw my toiletries in the duffel and quickly put everything else in the boxes downstairs. I sweep floors, wipe off counters, scrub bathrooms, feed a cat I don't like, leave a gift (three French linen guests towels) and turn the locks as I close the doors behind me.

I haven't decided where to go next, so I drive around the city for hours, stopping for coffee, then lunch. For the rest of the day I check the rearview mirror even more often than usual—if it's possible, that is, to do something more often than constantly.

Looking backward more than forward as I drive, I see no one following me. It doesn't necessarily mean that no one is.

Chapter 54

These days, when I go to the police to report another incident or to ask for guidance, I hyperventilate at some point on the way to the station. Each time I think that it can't *possibly* happen again, but I'm always wrong. One day I go to the police department to pick up a letter that will allow me to get a credit card under a different name. The Sex Crimes Unit is on the lower level of City Hall. I park in the city's underground parking ramp. It has twenty-four-hour surveillance and is connected by tunnel to the police department.

When I walk through the tunnels leading to City Hall, I pass the Domestic Abuse Center. When I applied for a restraining order a few months earlier, I spent hours in this center. I remember the day I discovered I'd have to go to court myself to obtain the restraining order. I knew instinctively that was exactly what Paul wanted, an opportunity to dare me in public to confront him, an opportunity to force me to spend money on legal fees to defend myself against him. I was convinced that he'd leave the courtroom empowered, safe from solid proof, emboldened to use even more sinister means to get to me. I ended up not pursuing a restraining order. The police didn't disagree with my decision. Sometimes it gets worse after you get one, they confirmed.

And here's the thing about your stalker, a detective pointed out that day. Even if he were caught, even if he were convicted and imprisoned, which is highly unlikely, he added, he'd be even more dangerous behind bars. "If you think he knows dangerous people now," he said, "they won't compare to the people he can get to do things for him in prison."

Within steps of passing the center this afternoon, I begin to hyperventilate. I push past my panic and take in enormous gulps of air, one after another. I ignore the stares of everyone I pass. Being embarrassed is the least of my worries. I stop before the door of the Sex Crimes Unit and take a gulp of nasty-tasting city water from the fountain.

A few minutes later, I stand in the reception area holding the letter the sergeant left for me in case she wasn't there when I came. But she *is* there. As I wait for her to walk to the front of the offices, I read the letter, overwhelmed with gratitude that such a compassionate professional had been assigned to my case. We stand and talk for a few moments, and then I leave, to experience the same terrified moments that I face every time I come to this place.

Wherever I park, whatever tunnel I arrive in, when I leave I inevitably take the wrong exit. Disoriented and confused, I feel like a frightened, abandoned child who can't find her way home, but I tell myself the feeling will pass and this moment is out of sync with the rest of my life.

Today I walk fast, even though I know I'm going in the wrong direction. Though I realize it's irrational, I'm convinced that stopping—even if only for a moment—to reorient myself would be a mistake. Ahead, on the wall of the tunnel, is a large red alarm button labeled "Press for Assistance." I'm tempted. I know my situation is not what they had in mind when they installed it, but newly learning as an adult to ask for help, I'm sorely tempted to ask for it now. I make a U-turn and retrace my steps. I find the stairs to the street-level entrance of the building, slip into the revolving door, and walk outside into cool air. Greeting me are sheets of rain darkening the solid granite blocks that make up City Hall.

I'm surrounded by people leaving work, shopping, heading for home. Everyone I see is either huddled in a doorway or dodging the rain under an umbrella or a hood. Here I am, lifting my head back to let the rainwater wash my face of fear. Water drips off my hair, onto my shoulders. I'm sure my face looks like the face of a woman just let out of the psych ward. I don't care, because suddenly I remember where I left my car, and the panic subsides.

I walk to the outside door of the garage, open it, and head for my car, which will be right where I left it, unharmed, because it's in a section under constant surveillance. I'll be all right, as soon as I'm driving away from the scene where I report crimes borne of another person's craziness.

Fortunately, my next stop is my therapist's office. These panic attacks render me nearly unable to focus, much less function, which scares me more than anything. She has diagnosed posttraumatic stress disorder.

I trust her judgment, but I find it hard to accept, or justify. I haven't been in a war. I haven't been raped. I haven't been physically attacked or molested. I haven't seen something so horrible it haunts my waking and sleeping dreams. I don't think my trauma is enough to warrant the symptoms I have. But when I do my own research, I discover I rate high in the symptom checklist.

Despite the fact that my nature is to discount rather than exaggerate, I have to concede that if there were a contest, I could be on the shortlist for the PTSD poster child. Even so, I'm ashamed and embarrassed by the diagnosis.

What are years of being stalked, I ask my therapist, compared to being in Vietnam, like my brother? She draws the distinction for me. When you fight in a war, she says, it's horrible, but you leave it behind you. You take the memories home with you, but the war doesn't actually continue once you're home. When you're being stalked, the terror repeats itself wherever you are—in your bedroom, in your telephone, in your computer, in every restaurant and shop and parking lot. And, she continues, it could be in the eyes of every stranger you meet or pass. You keep asking yourself: Are they following me? Are they watching me? Are they staking out everyone I love?

When you're being stalked, new terrors pile on top of the memories of previous terror, and, she adds, unlike a war, you spend an enormous amount of time and energy convincing yourself that what's happening to you is real. And that it's terrible.

Chapter 55

The idea of a gun is tempting. Some of the detectives advise me to carry one. Of course, take lessons first, they say, or do some practice shooting. This surprises me, but it indicates their assessment of the danger I'm in.

Each time someone makes the suggestion, I reply with a joke. I explain that sometimes I'm so focused on my work that I find my phone in the refrigerator and a carton of yogurt on my desk. So, I say, someone who can be this preoccupied can't be trusted to have a gun. In truth, that *is* a factor. I'm not sure I can be trusted with a gun. And I wonder where I'd keep it. In my backpack, I suppose, and under my pillow next to my cell phone at night.

My oldest brother, Pat, an ex-Marine, helps me put the question to rest. "I have one piece of advice to give you," he says. "Get a gun only if you're one hundred percent positive you could kill someone with it."

If that's the measure, I know the answer. If threatened, I'm sure I could shoot to maim or wound, but I'm not positive about the killing part. Besides, if I have a gun it will never be Paul in the line of fire, just someone else his money has bought.

Oh, don't mistake me. I'm plenty angry, but I have a rich fantasy life, especially in my sleep. In my dreams, I want to kill him, at the very least see him dead. In my waking hours, I try simply to wish this madness would cease. It worries me a little that I can feel my full rage only when I'm asleep.

We're alone in a room, his old bedroom, emptied of furniture except for the straight-back chair he's sitting on. I'm across the room with a gun, and we both know I could kill him that moment, even though he isn't tied up. I shoot, but shatter the wall instead of his head. Just to scare him. It occurs to me that I don't have to kill him because he'll undo himself. As I turn to leave the room, outside his window I see a woman, hovering like a mad hornet on steroids and dying to get at him. She has metal shields where breasts are meant to be and arms that could easily climb rocks without ropes. She looks as fierce as his past.

"Someone's here for you," I say, and smile. I open the French doors to the backyard and drift away.

This becomes my favorite dream.

I understand women who snap and kill men to save themselves and their children. I don't condone it, of course, but I understand the impulse. Eventually these women know only one thing for sure: their men would rather destroy them than let them live. The irony is that men like this—weak but violent men—can't stand being left. So leaving can make it worse. In my case, it turned into stalking. Even though we now have stalking laws on the books, often it's only when a woman dies that the crime is taken seriously—unless you're a celebrity.

With Paul, I became one of those women you read about. "I didn't know what he was like. I just didn't know." That's what we say. And it's often true, even when, in retrospect, it seems hard to believe. There's a certain kind of man who changes once he's drawn a woman in. We know in the end, but not at first.

Chapter 56

"Have you ever thought that you could be making this more fun for him?" asks one of the detectives. It's a sunny afternoon on the first warm day of spring, but in this belowground, windowless conference room, it might as well still be the dead of winter.

"*Fun* for him?" I say, momentarily stunned, and feeling dense, which I hate.

By being so difficult to get to, he explains, by being so quick to react. He suggests this might make it more of a challenge for Paul. "It's like a game for him," he says, "and that will increase the fun. The harder you are to get, the greater his satisfaction when he's successful."

I know he's right. I hadn't thought of it before, but all I have to do is look at how I feel every time Paul doesn't "get me." Aside from relief, I also feel satisfied: at least some of the time, I'm resourceful and quick enough to see him coming and move aside so his rage merely ricochets off the shadow of my life.

Yet this is not a game for me, so if *I* feel this satisfaction at staying ahead of him, his pleasure must be enormous when he gets a direct hit.

These detectives know more than I do about sex crimes and

about stalkers. At the same time, they regularly remind me that I know my stalker better than they do and that my instincts about him usually prove correct. My sanity depends on a fine balancing act: listening to their advice and trusting my instincts. It doesn't leave much room for any other life. I imagine this is just what Paul intends.

I compromise on their suggestion that I settle down in one place for a while. I don't intend to make this fun for Paul, but neither do I intend to make it easy. I decide to spend the summer split between my oldest brother's house in the suburbs and my mother's house a few hours away.

I feel safest of all when I'm at my mother's house. I know it's a false sense of security. Besides, I can't stay here forever; I need to be in the city to keep my business going. My brother's house provides a different kind of safety in the constant activity that swirls around teenage children.

Pat doesn't seem entirely convinced the stalking is as bad as I say it is, and I see his point: the worst happens when I'm on my own, so no one else sees everything firsthand. I'm hurt my brother doubts the severity of my situation, but I let it go. After all, he does the brotherly thing. I'm sensitive to the danger for them, especially their children, but they push that aside and show me only kindness.

The first night I'm in their house, all the kids are home, so I sleep on the couch. Just before he goes to bed, my brother comes into the living room with Keisha, their sweet-faced Akita. She usually sleeps beside their youngest child's bed. "Now, Keisha," my brother starts, and Keisha moves to his side and fixes her eyes on him. "Kate's the one who needs the most protection now, so when she's here, sleep beside her, wherever she is." When Pat leaves the room, Keisha's eyes shift to me, and she drops to the floor next to the couch.

I settle in with a John Sanford novel to remind myself how much worse life could be. I let one arm drop down to stroke Keisha's thick black-and-white coat. From that night on, whenever I'm in their house, no matter how long it's been since my last visit, no matter what bed I sleep in, as the lights of the house go out, Keisha finds me.

Chapter 57

When it comes to my field of friends, my being stalked proves an easy way to separate out the chaff. Ironically, given what I've heard from others, the police need less convincing than some of my friends and family. In fact, some of my siblings and friends dismiss my situation altogether. Worse yet, some feed off it.

One friend wants to hear every detail of the stalking, but she refuses to keep them to herself, despite my pleas. Her closest friend has a connection with Paul, and when I find out my friend has been entertaining her friend with my story, I cut off contact completely. Just like that. I'm becoming well practiced at leaving people without looking back. While that may seem unkind in ordinary circumstances, it registers with me only as lifesaving now.

I can easily count the number of people who know where I live. It's a short list, carefully chosen. Even some of those closest to me would rather not know. Greg, for instance, a strong, six-foot-tall military type, laughs and says: "Don't tell me. I'd break if anyone tried to torture it out of me." We laugh, but the truth is that it's a burden for people to have to remember to keep my secrets. Curiously, I feel less alone than I felt when I had more friends.

A few loyal pals do exactly as I ask, meeting me wherever I feel safe and protecting every nugget of personal information they have. These friends share a fantasy, and they argue for the privilege of being the one to appear at Paul's door. "She understands now," the designated negotiator would say. "She wouldn't take your money before, and you need your women to take it. *So*, how much money does she need to take for you to leave her alone?" Then, the appointed one would smile sweetly and stretch out a hand for a check.

"You can turn around and give it to charity," they tell me.

When I meet my friend Phil for lunch one day, he says, in a matter-of-fact voice, "You amaze me."

When I ask him why, he says, "You haven't transferred your hatred of this one man onto any of the rest of us."

"But he's only one man," I say. "And think of all the men I know—men like you—who are loving and kind. It would be wrong of me to brush others with his insanity."

"But understandable," he laughs. "Besides, we all have our demons."

"Yes," I say, "but some of us face them and choose to live as much in the light as we can manage."

It's as simple as that. We all have our demons. And we have a lifetime of choices, to harm others or not. I don't fool myself: I'm dealing with a man whose demons are enormous and who's too lazy to do the work required to make a psychological breakthrough. But I also never forget: he's only one man.

I have one particularly loyal client without whom my business would have failed completely. Since Paul infiltrates every area of my life, it's not safe to take on new clients with whom I have no prior relationship. I've already had one experience of nearly agreeing to do a freelance job for someone Paul paid. Luckily, I saw through the offer and walked away.

Yet no matter how careful I am, no matter what I do, things don't improve. People watch me from parked cars. I'm followed. I get more than my fair share of hang-ups on the phone. And messengers try to deliver packages to my door or to my mailbox center from people I don't know. I never accept them.

For a while now, the police have suggested that a private detective might be able to help me in ways they cannot. When pressed, one of the detectives on my case recommends someone, unofficially. I've resisted taking this route so far because of money. Staying on the run takes time and energy, and I'm so tired from all the stress, I can't work full days and weeks like I used to.

So I make less money, while I spend more to stay safe. He'd love that.

Chapter 58

Our first meeting takes place on a hot afternoon in August. The time has been set, but the location left open; I am to contact him when I arrive at a place where I feel safe and am sure I haven't been followed. He'll meet me there as soon as I call. Not on your cell phone, he says, not on any cell phone, he repeats, as if English is not my first language. I ask him how I'll recognize him. "You'll know me," he says. He knows I did my homework, reading about him in the newspaper archives.

His walk is what I notice first when he arrives at the coffeehouse I've chosen—the way one leg is always late for its gait. Then, of course, there's his face, caved in on the left side, with the eyebrow raised, caught by stitches in an eternal look of query. He looks smart and wise. He used to be a state trooper—before he was shot and left for dead on the side of a highway. He'd stopped a speeder, stumbling on a drug runner on his way to Canada. He's brought his partner, a Native American woman, who looks as if she knows how to take care of herself and anyone else in the room. She also looks kind. I decide I like them before we even shake hands.

A faded green awning juts off the low brick building and

creates a private corner for us to talk. He asks me to start from the beginning and tell them the facts. "First tell us a little about yourself, then tell us everything you can about this guy and about what's happened to you," he says.

Half an hour later, as I finish, I say, "I guess you'd say those are the facts," though I know I've embellished my narrative with plenty of opinions. I take a deep breath and add, "Here's what I think: the timing of our relationship, dating just when his father was murdered and his family's secrets were exposed, means that I've accumulated enough of Paul's secrets to make me dangerous to him, and, much as I'd like to, I can't give them back. Ever hear the Berber proverb about getting between the fingernail and the skin?" I ask. "That's how I must feel to Paul."

The private detective says, "I misunderstood. I thought you said you were a writer."

"I am," I say.

"*And* a psychologist," he says. It's not a question.

"No, just a writer."

He laughs. "Well, you *sound* like a psychologist. You've really got this guy pegged, and it's the kind of psychological description that fits what's going on."

"I read a lot," I say.

"I'd say so," says the ex-trooper's partner.

"It helps me cope," I explain, "understanding as much as I can. It's my way. Knowing as much about a thing as I'm able."

We laugh, which relaxes the tension of the story that up until now felt like a burden I keep handing over, bit by bloody bit.

"Being able to figure this guy out helps you," he says.

"I'm counting on it," I say, thinking I'd better be right about this. They'll talk to the detectives on the case, they say, as soon as I give the police permission. "Then we'll tell you what we think you should do," says the one-time trooper.

One of the first things they plan to do is to check Paul's garbage to see what they can learn about his life and habits. At first I'm a bit disturbed by the idea, but quickly realize I'm happy to pay for such a thing.

A few days later, I get an update. They've talked to the detectives, watched his house, and gone through the garbage he put out the night before. "Amazing what you can tell about a person from his trash, even one as secretive as your guy," he says. "Now that you've gone to the police, he's apt to be even madder than before. Not scared. The easy ones scare. They can't stand the pressure of the police watching them. Your guy doesn't scare too easily." He stops, but I don't break in. "Guys like this are really cowards," he explains, "but your guy's too arrogant to scare. If he was *really* smart, he'd be plenty scared. His kind of life eventually catches up to a person."

I stop him now. "Don't ever call him 'my guy' again." I try not to sound harsh, but I don't succeed, and I instantly regret it. I rush out an apology. I need *this* guy on my side. I'm way over my head in trouble, and the only way I can see out involves turning my life over to strangers such as these. They'd better be good, I think, and they'd better be right.

And they *are* good, as I see in the following weeks as they offer up bits of information about Paul that depart from his story. They find evidence of a fourth marriage. They tell me he has at least one extra house no one knows about. Paul's life is based on more layers of lies than I imagined. The expression "master of deceit" becomes personal.

The private detectives have been helpful. They've found out that Paul had planned to buy the house he rented across from me until he discovered I was moving, which further validates that his only reason to move to my neighborhood was to frighten me. "This is one scary bastard," the woman says. Digging into his life the way they do, questioning acquaintances, spreads the word that he's under suspicion. It also lets Paul know that I'm surrounding myself with professionals.

They don't charge much. "It's what we can do to help you," the detective says. But still, I can't afford to keep them around me forever. Tempting as it is.

All I know is that, somehow, I have to stay one step ahead.

Chapter 59

After a year of being on the move, I'm ready to try one of the other options: living in one place under an assumed name. Through a friend who owns real estate in several states, I sign a short-term lease under a false name. He arranges what would be impossible to arrange otherwise: the town-house manager thinks I'm one person, the electric company thinks I'm another, and the people who pick up my garbage think I'm yet someone else. Only my friend and his assistant know who's really renting the town house, which is tucked behind woods, reachable only by one main road and a dead-end driveway. You'd have to be on private property to watch this house.

When I move in, I take a few things out of storage: a rocking chair, a couple of lamps, my desk chair, the gateleg teak table I'll use as a desk, and a futon. I find the box I packed with my down comforter, pillows, and linen sheets, and the one with kitchen essentials: electric kettle, teapot, French press, an omelette pan and a few bowls, mugs, and plates. Most important of all, I bring Keisha with me. My brother and sister-in-law offer her for as long as I need.

I'm anxious to have a place of my own again, but for the first time I'm afraid to live alone.

Fear, especially of not knowing enough, drives me. My own therapist admits that most psychologists don't know much first-hand about what it's like to be stalked. So I go to a conference on stalking sponsored by a professional association. At the end of the morning, I tell the main presenter I'm being stalked and would like to consult with the country's most knowledgeable expert.

"I want that to be my specialty," she says, "so even though I'm new at it, I'd be happy to consult with you."

"Thank you," I say in my most polite voice, "but I can't afford to be on anyone's learning curve. Who's the best now?" I keep smiling even though I'd like to give her a verbal slap for taking this so lightly.

She gives me the name and number of a forensic psychologist in California. I call to arrange for a consultation in February.

First, however, I move into another town house owned by my friend's company. This one offers a much more private setup than the unit I've been renting for the past four months, which will make it even easier to keep track of strangers. I sign a year's lease.

I've chosen my new town home carefully. It's a second-floor unit with no windows in front. The back wall is nearly all windows, opening to a small woods of maple and elm and oak. Natural linen curtains let light in and keep eyes out. Anyone watching the house from the back is in full view.

Besides, Paul wouldn't think to look for me here in the suburbs. He knows I prefer the city or the country. But even in this home that knows me by another name, I settle tentatively, because I expect him to find me eventually. I can't help but think of Anne Brontë's words in *Wildfell Hall.* When Helen leaves her husband and arrives at her hiding place, she says, "But for one disturbing care, the haunting dread of discovery, I am comfortably settled in my new home."

That's how I feel. This is far better than the days I leaped from home to home, stopping in rooms just long enough to plot my next move, when I traveled endlessly because I was too nervous to stay in one place for long.

Since the November day I walked out of my apartment and checked into a hotel, this will be the sixteenth move I've made in as many months.

Chapter 60

Sunny California: my first visit to the sunshine state, and it's raining when I arrive. Actually, I'm relieved. I won't feel so much like an alien in the land of bare, tanned skin. I've come on a Saturday so I can acclimate myself to the city before my Monday appointment with the forensic psychologist. I'm here to settle the question I can't put to rest: How dangerous *is* Paul?

Signing my fake name with a credit card to back it up, I check into a small hotel in San Diego's Lamplight District, a neighborhood almost fashionable again, and set off on the tram to find art.

In a small museum in Balboa Park is a Pissarro: a snowy street in a village, two shadowy figures in the background, and a shawled woman in the foreground. This scene of winter quiet is shot through with questions. Does this woman live in this village or is she just passing through? Either way, is she safe? And the men approaching her: Do they mean her harm? The painting reminds me of a Monet I saw a little over a year ago: a woman with a red shawl and a face sadder than the sunless winter sky above her.

Much as I try, I can't stop thinking about Paul. I project the possibility of sinister intent even to this cool Pissarro. I'm angry.

Who in the hell has to travel halfway across the country to get a risk assessment? Nobody I know. I spend my life and time and money doing things I would have once considered bizarre. Now such things constitute my everyday life.

I'm ready for this forensic psychologist to tell me I'm wrong to think Paul is dangerous. He's a nuisance, an annoyance, she might say, but he's not really dangerous. Relax, get on with your life.

But she doesn't say anything close to this. Instead, she listens to my story during the first two sessions, and talks to the detectives in between. In our third session, she asks about the precautions I've taken. They all sound appropriate given who he is, she says, but you haven't gone far enough. Then she describes how she sees the situation: my stalker is a persistent, patient, relentless man. A long-time stalker, she says. He has the money and the resources to keep himself removed from the danger as well as the personal consequences of his obsession. As long as something meets his need, she says, as long as it fulfills his vision of himself as a persecuted and underappreciated man, then he believes it's all right. People like this don't view right and wrong the way others do. She reminds me that the majority of stalkers are never charged, much less convicted and imprisoned. She suspects that I'll be his target until he does what many stalkers do: shift his obsession from one target to another.

"But I don't want the end of my stalking to be merely another target's beginning," I say.

"It's not up to you," she responds. At our last appointment, she details measures even more secret than I've planned. What she describes is an undercover life, one I've seen only in movies. "When the pressure around you is to relax your vigilance and your security measures," she says at our last meeting, "don't do it.

This will probably last for several years. He's the kind who'll leave you alone for a while, just long enough for you to drop your fear and relax your safety precautions. He'll start all over again. And he'll probably do that a number of times. It's a game for him."

I tell her what makes me angriest: "I'm afraid he's stolen my ability to be fearless. I don't think I'll ever get that back." Until now, I've held back tears. Now, they come.

She smiles, which unsettles me, until her next words sink in. Actually, she says, she doesn't think I've ever had a healthy amount of fear. Everyone needs to have *some* fear. And now that I do, I shouldn't ever go back to being fearless. Having some fear can keep me safer.

When I get up to leave her office, she says, "Do your best to get on with your life."

I feel as if she's just said: Stand in the middle of the freeway and read a book. Live this way and be serene? I can't imagine how to do this. I tell her so.

"You must always be vigilant," she says. "You must never let down your guard. It's wise to trust almost no one. But," she adds, "you mustn't let him destroy you or your capacity for joy and love."

I don't know how to do this, I say.

"You'll find a way," she says. "You're strong and smart and brave. And you've learned how to ask for help. You'll figure it out."

Until this moment, I thought I'd lost the traits I treasure most, but now I see I mistook fear for cowardice, and I confused asking for help with being weak. It's as if I've been holding on to courage the whole time and only now look down and see a perfect sphere of it in the palm of my hand.

I fly home to learn how to hold vigilance and freedom side by side. I expect it to be the longest learning curve of my life.

Chapter 61

"Bitch." That's all it says. The note tucked into my windshield wipers is not in his hand, of course, but it looks like his particular *way* of printing—an extra taunt. I've run an errand at a large shopping mall on the opposite end of the city from my house. As the police advise, I rarely take the same route twice. Yet I return to my car to find this note.

Two days later, I drive to a client's for a meeting. I have my own office and voice mail there. When I check for messages, the first one, in a man's clearly disguised voice—not Paul's, of course—says: "Bitch, I know you're there. And I'm going to get you." Or "kill you." It's not clear. I am, however, quite sure I recognize the voice of the woman laughing in the background. She's young and gullible, sweet-looking but malicious, and always short on cash, just the kind of woman Paul could easily engage in the hunt. I know I can't prove it's her, but I won't forget. I tell the president of the company about the call. We consult with the human resources vice president and the corporate attorney, and they insist I call the police.

I phone the local precinct where my client's offices are located, and when the detective shows up, he's quite casual about the call.

Hard to tell what he's saying at the end, he says after listening to the message a few times.

I explain that the detectives in the Sex Crimes Unit downtown want to be called so they can brief other police departments I deal with.

No need for that, he says.

"This has been going on for years. They know all about it," I explain.

"Just tell me about this one incident," he says. "That's all I care about."

"There's not much to tell," I say. "Just the message, but it matters that it's part of a pattern."

"Tell me about your relationship," he says.

"Look, I lived with him and left him, and he's been harassing and stalking me for years," I say.

"He must really love you to be this persistent," he says.

"Did that message sound like love?" I ask.

For some guys, it is, he says.

I hold my temper, but not my tongue. "I gather you're a trained detective," I say, "but you must not be familiar with the crime of stalking."

Enough, he says.

"Not enough," I reply, "to realize what a stupid remark that was. This is *never* about love," I say. "It's always about power and control."

"I don't know about that," he says. "Looks to me like this guy must really love you."

"Well, then, there's nothing you can do to help me," I say.

It strikes me that I've been spoiled by the help I received from the Sex Crimes detectives. I try to convince myself that this man's response will be an exception, not a rule. But mostly I'm losing hope.

For weeks after I receive this message, each time I drive in my car alone, a single image catches my fear, like a flash flood after sudden rain. A window is lowered in a passing car, an arm is raised, a gun is aimed, and a single bullet enters my head.

Perhaps that's how he'd have it done. And just like that, it would be over, in the most ordinary moment of a life.

I want to get out of town, have a break from work, and from the insane way I have to live. I can't afford to go far, so I call my mother and ask if she'd like to take a trip along the Mississippi. We leave on a twelve-day drive the morning of my forty-eighth birthday.

I've stood at the very start of it, canoed down part of it, and flown to the Louisiana banks so I could dip my hand in waters that dump into the Gulf of Mexico. This trip will bring us through five states that border the river.

We have no reservations, just the books I've bought that will give us ideas about what to see and where to stay. I've mapped out the general route, as close to the river as possible, and sometimes that means weaving our way across the river, state to state, several times a day. We travel well together: we both enjoy having plenty of information but no fixed itinerary.

For nearly two weeks I live without fear.

Chapter 62

I'd make a good hermit if I didn't have to make a living. Some days, I sit in the shelter of my deck in a white mesh rocking chair screened by tall plants. All I hope for is that no one will cross my narrow path of sight. My notebook is in my lap, and my pen is in my hand. I write a poem about battered crows whose coagulated brains are being suckled by greedy bluebottles, and vultures that swoop low. It's not a good poem, but it does the trick.

I've got so much anger, it's no wonder I feel crowded in any other place but my own. I'm most content these days when left alone, with my books, my pen, myself, to do client work and write angry poems.

At the end of one such day, I stand in my kitchen and prepare sauce for pasta. I slowly and methodically cut up raw vegetables, and somewhere between the onions and the peppers I start to feel better. Chop. Chop. Chop. The rhythm calms me. I sauté the vegetables in olive oil, and watch them change color and shape. I squeeze Italian tomato paste from a tube, shred fresh basil, and add a hint of oregano. I set the mixture to simmer.

While the pasta boils, I read the paper. A news junkie, I've learned not to mind reading day-old, even week-old newspapers.

No mail at home and no newspaper delivery either. Too easy to track. I wonder if I'll ever again have the kind of life in which I wake up, walk to the front door, and find the daily newspaper waiting for me.

I feel restless, so I busy myself with a colander, a bowl, and spaghetti tongs. I reach for two glasses, one for water, one for wine. Communion. I stand and wait at the altar of hunger.

My kitchen is in a corner of the great room. Nearly everything is covered in calming shades of cream and taupe: the couch, the carpet, the walls, the countertops. The role of color is delegated to dishes, bookends, and, of course, paintings. Over the fireplace, my swans are headed for the woods. I put on a Mary Black CD, the one with the song "Paper Friends" (I play that one over and over) and flip the switch on the gas fireplace. With a bowl of pasta in my lap, I sit and watch the approaching night outside my windows.

An hour later, I clean up the kitchen, switch off the fire, and close the windows. Then I check to make sure the doors are locked and the alarm system is on. I turn off the lights and make my way to my bedroom. As I lie in the dark, the shadows of the room bring to mind a single line of dialogue from a movie whose name I don't recall: "All monsters look normal until they're caught."

I can't actually picture myself checking every door, so I get back up and walk down the hall. I won't be able to sleep until I'm sure, so I check the front door, the garage door, and the door to the deck. I open the alarm panel to make sure it's set correctly.

As usual, under the spare pillow on my bed is my cell phone. I've added an ice crusher: a round stainless-steel disc on the end of a long, flexible, white plastic handle. If I hit someone right between the eyes, it could knock him out cold.

Chapter 63

I watch the garage door close. I always sit in the car until the bottom meets the cement and wait the extra seconds it takes to make sure it doesn't bounce back up. Only then do I drive away.

When I leave, I wave to the neighbors who are in our shared driveway washing their car. We've had the occasional conversation and glass of wine, but mostly I keep my distance. I'm not comfortable with my duplicity: they know me by a false name and a made-up backstory. It's necessary for my safety, but I don't like to deceive good-hearted people.

It's a Saturday afternoon, and I have an appointment with my therapist. I'm ashamed of my fear, I tell her, and explain how the dark rushes in when I close my eyes. One minute I look fine. I'm functioning. I'm barely remembering. Then come the memories and the whole-body brownout: my hearing dims, my vision freezes, and my connection to anything outside myself frays. By nature able to hyper-focus, I suddenly can hardly focus at all, much less function. Each episode lasts for minutes. But the fallout—the hyper-vigilant senses—lasts anywhere from hours to days.

I'll hear and see and smell more acutely. Even my skin feels as if it's snapped to attention. And I'll be in full-flight mode even more than usual, moving too quickly, bumping into furniture, bruising myself.

I loathe this readiness for battle, and since it's an invisible one, I often feel foolish. When I tell my therapist how afraid I am that I'll never be able to function as well as I used to, she smiles and says: "Because you usually function at such a high level, you've dropped to what most people see as normal. This isn't forever," she adds. "It'll be all right."

When I return home, the overhead garage door is up and the inside door between the garage and house slightly ajar. My neighbors appear as soon as I drive up.

"We've been listening for you," she says.

About an hour after I'd left, they heard the alarm in my house go off. They went to my front door—also open about a foot—and called for me. Getting no response, they decided to wait.

They offer to come through the house with me to make sure no one's inside.

I tell them about the stalking. "I don't expect you to help me," I say.

They insist on coming in anyway, so together we walk through the house. Everything is as it should be, except for the open doors. It's the pattern I've grown accustomed to: nothing concrete, no "evidence." Yet, once again, a sense of security is ripped from my hands.

I no longer have the heart to convince a new set of detectives. It would be a different station, an entirely fresh cast. It wears me out to keep explaining. Of course, each new police officer needs to hear the whole story, but it's nearly as bad as living through it the first time.

I'm exhausted by it all. I've lost track of how many times I've come home to an open door I know I locked. Maybe five, maybe six, maybe ten. As for power to my house being cut: maybe three times, maybe more. I quit counting the number of times my phone has been tampered with a long time ago. All I know for sure is that it's more than I imagine.

To make it worse, I've just discovered that a friend of Paul's has been named Commissioner of Public Safety for the state. That means he has jurisdiction over the Bureau of Criminal Apprehension and the State Patrol and God knows who else. For the first time, it occurs to me that going to police I don't know could actually put me at greater risk. Paul would find it easy to convince police in this suburb that I'm the one who's dangerous, that I'm the one to be watched. He seems to have eyes and hands everywhere—to watch me, to pick locks, and to hack into computers. I wonder how far Paul's arms reach, and whose pockets would be immune to his "generosity."

And no matter how helpful the city detectives have been (and I trust them completely), they can't do anything for me unless things happen within the city limits. Nor can they help me if Paul finds someone else in law enforcement who would rather harm me than help me.

It strikes me with new force: I'm on my own. There's no stopping him.

Chapter 64

I feel like a fraud. After all this moving, I think I've learned to redefine home as an internal state, but each new prospect of travel makes me homesick even before I pack, and not just for people. When I leave, I grieve the flash of blue jays through the branches behind my town house, the way maple leaves flicker like a thousand fans in a slight breeze, the look of the moon from a certain angle. And I yearn for the bed that knows the shape of my frame and worn linen sheets that cradle me in sleep.

Most of all I miss the trees I've come to envy, for they do what trees are meant to do: stand in one place, swaying but not moving. I make myself travel so I won't give in to my fears. It's the only way I feel as if I haven't been robbed of my entire way of being. Mom and I have planned a trip to Europe. We spend a week in London, then visit my friends in Yorkshire.

In part, we're researching her family, in particular the grandparents who met and married near Durham (one having come from Ireland, the other from Scotland). So between scavenger hunts in the London archives and county record offices, we visit villages where her grandparents lived and worked before immigrating to America.

Midway through, our trip takes an unexpected turn. What Mom took to be bronchitis turns out to be heart failure, which puts her in Edinburgh's Royal Infirmary. Instead of two weeks in France, we spend four more weeks in Scotland. She's got "leaky" valves, and they need to stabilize her so she can get home for surgery. On days when she's nervous and worn out from tests, I bring in all the newspapers published in London and Edinburgh. It keeps us occupied for hours. As does our conjecture about her ward mates, especially the one Mom nicknames Sparky. Every couple of hours the wiry woman with the spiky red hair puts on a jacket and goes out to the fire escape for a smoke. She fascinates us.

On days Mom feels strong, I leave the hospital for a couple of hours to take long walks, sometimes across Castle Gardens, sometimes up to Waverly Place, but more often than not to a museum or gallery. One afternoon I walk to the National Gallery, a centuries-old building in the middle of ancient streets. Not looking for anything in particular, I'm struck by a small painting: a portrait of a woman who sits, in a simple gray dress broken by bits of white—a bib, two cuffs at her elbows, and a close-fitting bonnet. She looks all the more plain in the shadow of a portrait of Madame de Pompadour, a study in pearls and lavender and blue.

The records leave much to be desired, but most accounts say that Sarah Malcolm went to bed one night more than two hundred years ago and woke the next day to kill her mistress, Lydia Duncombe, and two fellow servants. Then, once the deed was done, with the memory of blood pounding at her heels, she posed for William Hogarth in London's Newgate Prison before walking to her execution at the age of twenty-five. Her neck is taut, her cheeks flushed. Her right eyebrow, just the one, is arched, as if to dare her audience to pity her. Held before a prim bodice, the rosary in her hands curls like a snake of beads on the table before her. She doesn't look like a guilty woman beckoned by the rope of death. But then again, killers can look as ordinary as the rest of us, I suppose.

When I get back to the hospital, I can tell Mom's got something on her mind. When I ask her if something's happened, she tells me no, but that she does have something to say and wants me to listen carefully. "If anything happens to me here, you're not to feel badly. No one could have taken better care of me."

"You're going to be fine," I say.

"Yes, but if not, remember what I said. I've had a good life, and it will be all right if I go."

Once again, I'm the one who can't keep the tears in.

"I mean it," she says.

"Thanks for telling me," I say, but then I start to laugh. Suddenly all I can think of is my seven siblings (and everyone they bring with them) lining up to greet me as I arrive home—with no mother.

"And what's so funny?" she asks.

I describe the image. "I'll tell you what's going to happen," I say, "if you don't make it out of here. I'm having you cremated, and I'm taking you with me. If you don't make it, I won't be going home either!"

"Be sure to take me to Switzerland," she says, laughing and playing along. "Your father and I had a wonderful time there, and I've always wanted to go back."

That night, when I go back to the hotel, a Georgian town house that overlooks Dean Village, for the first time since she was hospitalized, I don't have the energy to address the pile of phone messages from my siblings. I'd called my oldest brother earlier in the day and asked him to spread the word. They act as if I'm holding something back from them. But they know all I know: at least for today, Mom is okay.

Chapter 65

In early November we fly home, and the first night back I realize I'm not glad to be home at all. I've been spoiled by feeling free. As it turns out, I'm not home for long anyway. Mom has open-heart surgery a few weeks later, and for nearly a month I stay in the hotel wing of the hospital. She nearly dies before getting to surgery, and, for more than two weeks afterward, she's in intensive care. We're not at all sure she'll make it at first. We take shifts. There's not a moment when at least one of her children isn't either in her room or in the waiting room right outside her door. Usually there's more than one, along with spouses and grandchildren.

A month later, Liz and I take Mom home. We make a pact: together, we'll look after her as long as she needs us. I move into Mom's spare bedroom. Liz, who lives a few miles from Mom's house, arranges her family's schedule around Mom's needs. Besides keeping her company, we take her vitals, keep in contact with doctors, count out pills, help her bathe, and make sure she gets nutritious meals. Mom recovers from her surgery. However, we've learned she has another failing valve, so if she has two more years, she'll be lucky. Mom doesn't want to know this, so we keep it to ourselves.

Every morning, I brew myself a single cup of coffee and make her a cup of Scottish Breakfast tea. As we sit across from each other at the breakfast table, I count out the day's pills. As she works to get them down, I read her a poem. Some days it's Dickinson, other days Grennan or Auden. I introduce her to Mary Oliver and often she asks me to read "Morning Poem."

Later in the day, while Mom rests in her chair in the living room, I get out my journal and sit on the couch across from her. Even when she naps, she has an open book in her lap, like a child unwilling to let go of a favorite plaything. I think I'll feel better if I can write about the stalking. But before I get too far, the post-traumatic stress symptoms start in. As frail as she is, Mom can tell by looking at me when I'm struggling. She asks me to tell her what's happening.

I explain that it's as if someone suddenly dimmed the lights and turned down the heat. Connecting with anything outside myself feels like an enormous effort. Even getting out of the chair and crossing the room seems too much. I don't know the biology behind this phenomenon, just the psychology, which gives me no comfort, only understanding. I can barely complete the bit of work I need to do for the few clients I can handle. It's all I can do to take care of Mom and make it through some days.

After she sees me go through this a few times, she says, "Maybe it's too soon to be writing about it. It'll be easier later." And then she suggests we watch a movie. We start in on *Pride and Prejudice* with Jennifer Ehle and Colin Firth, one of our favorites.

This classic story of ordinary people full of foibles works its magic: for hours, we're transported from our own woes into those of Austen's lovable characters.

* * *

While we're waiting at the hospital one afternoon for Mom to receive what are now monthly blood transfusions, Liz asks me if I ever hear anything about Paul. Sometimes news comes from unexpected sources, I tell her. The wife of one of his cousins, whom I'd run into in a bookstore, let slip that Paul has divorced again. I'd found out he graduated from medical school in a similar way. One day when a close friend was in her backyard watering plants, her neighbor crossed over to talk to her. The neighbor had just returned from her granddaughter's graduation from medical school. She was excited to show my friend the young woman's name on the program.

"Oh, let me see," my friend said as she reached for the program. As she pretended to admire the name of this young woman, my friend was actually searching for another name. And there it was: Paul had finally graduated from medical school, two years behind schedule.

Chapter 66

I'm staying with my mother at her house almost all the time now. She's getting weaker, which she sees as a setback rather than a sign. I want to stay with her so she can hang on to her life and her home as long as possible.

One night we watch a made-for-TV movie about a man being stalked by a woman. Our dialogue during the commercials is better than the movie. We discuss the rarity of the victim being a man and the stalker being a woman. The victim in the movie, a savvy-seeming professional man, falls apart after one threatening phone call. At the next commercial, in a voice full of disdain, Mom says, "He's certainly not handling this very well. This is *nothing* compared to what you go through."

Just after the show ends, my cell phone rings. It's the manager of my town-house complex. The front door of my house is open—not broken open, she says, just open. She knows I'm being stalked and that I'm out of town taking care of my mother.

I can hardly bear to drive the three and a half hours to see if anything has been taken. My mother is so disturbed, she offers to drive with me. By now, she can barely walk unaided, and while she eats hardly anything her body is greedy for the blood transfusions she receives every few days. I know she can't make the trip. She tries to convince me she can. She imagines her will is enough to carry her there. Small to begin with, and more frail every day, her love, by contrast, is enormous. We settle on my leaving the next morning for an overnight trip while she stays with Liz and her family.

"Stay in a nice hotel," she says as I pack my bag. "And order room service tonight. My treat." It's her way of trying to make this feel less awful than it is.

My cousin Maggie meets me at the town house so I don't have to go in alone. As usual, the break-in leaves me with no evidence of a crime, nothing to report to the police. Just an open door. On the way back to my mother's house, I try to imagine what it would be like to live in the small town where she lives. Living in a place where I'm well known is the only protective measure I haven't tried yet. But I'm not sure. This town, equidistant from the country's northern border and the drop-off lane of a major airport, seems too close to the end of the civilized world for my comfort, as if it might drop off the edge if they don't watch it closely enough.

A few days later, I sit in the corner of a coffee shop: a coffee shop so I can be alone to think, a corner so I can watch the door. I have the questions, and I need to come up with the answers. Can I live here and fill my days with writing while I watch my sister's children grow?

I've been here for nearly a year and a half, caring for my mother. Perhaps actually moving here is the next thing to try. I can already feel the irony of making such a move and having my mother not live long enough to benefit from it. One evening, my mother taps into my thoughts. We're sitting in her living room.

She's wrapped in her usual combination of cashmere and fleece to keep her shrinking self reasonably warm.

"Come here," she says, and stretches out her arm for me to take her hand. I sit on the stool in front of her chair. "Listen to me. Listen carefully." She holds my face in her hands and looks directly into my eyes. "I'll be with you always, wherever you are." And then she kisses my forehead. I lay my head in her lap and let her stroke my hair, as I've seen her do a thousand times, to her grandchildren, and her children before them.

The love behind her words is so clear it's almost palpable, both the affirmation and assurance that when a person loves you, you never quit carrying them with you. If you're deeply loved, such words have no expiration date. Some people, perhaps most, never know such love, so I know how fortunate I am to have these words, spoken in a gentle voice full of resolve, to bring me comfort for the rest of my life.

I also know the hatred of similar words, like Paul's, that have stayed in my bones. I can still feel him, leaning into my car, his hands reaching for my face. "You'll never forget me. I'll always be with you." The significance of his words has only increased over the years. He wanted to be important to our family. He's finally succeeded: he's at the center of our lives.

That night in bed I decide to give up my home in the city and move full time to this small town where strangers are obvious, and loiterers are asked to move on. Being known can be safer than being anonymous. I've tried it the other way for seven years, burying myself in a large city under a false name, constantly moving. That worked only for a while. Perhaps nothing will work forever.

A few days later, my mother asks me to take her to the lot I'm considering building on. As we circle the property, she says, "Yes, this will be a good place for you. I can feel it." I've already planned how to arrange the ground floor so she can live here, too, I tell her, but I don't dwell on it. She wants to think she'll be well enough to live on her own again. And who am I to take away such hope?

Within a month, just hours before thick snow blankets the late April morning, my mother dies. I've known this would happen soon. With the breaks between blood transfusions closing fast, I've known her days were dwindling. Yet somehow, when the moment arrives, it startles me. A few of us stand by her hospital bed while she eases her way out of this world. Just before she takes her last breath, her forehead creases, as if she, too, is surprised that the end has come. One by one, we say our good-byes. When it's my turn, I stroke her hands, which look so much like her mother's (and like mine). I lean over her, and, as she did to me so many times, I cup her face as if it were the most precious gift I'd ever been given. And, in fact, it was.

For beyond her words, which moor me to the feeling of being loved, she taught me the most enduring lesson of all: how to tuck home inside yourself. It's simple really. You take the memory of love and lodge it in your heart.

Chapter 67

Even in near-deafening grief, my decision to move sounds as clear as ever, so I have a town house built on the property. Then I buy a cemetery plot. Now that my parents are both dead, it seems irresponsible not to take care of the business of my own death. I draw up my will, sign a health-care directive, and purchase a gravesite at the foot of my parents'.

This all sounds matter-of-fact, even to me. But given what I've been through over the past seven years, it seems only practical to have my affairs in order. So I make the arrangements and don't allow myself to examine the significance too deeply. My body, of course, pays the price, in muscles that constantly scream for attention.

I move into my new home, install a high-end security system, and rent a mailbox nearby. I remember the advice of all the professionals I've consulted: Don't relax your vigilance. A break in the stalking doesn't mean it's the end. This could go on for years. Anything can trigger another rash of harassment: the mention of my name, a sighting, a random thought. I watch a program on television about stalking and hear an expert say: It isn't fair, but often the only way to get away from a stalker is for the victim to move. Ah, if my case were only that simple.

At least for now, I live in the middle of the woods, on the edge of a lake that requires me to cross the Mississippi River if I go into the nearest town. Just outside my back door is the reservoir. Once again, my home is near the river that's become the spine of my life.

Chapter 68

It's just after midnight on New Year's Day. I've greeted the year quietly, far from any crowd: tea with one friend, dinner with another. My New Year's wishes, still modest, haven't changed: safety, work, contentment. I'm asleep by one o'clock.

The alarm in my house, set high enough to wake any neighbor with normal hearing, blasts me from my bed. I glance at the digital clock: 1:33. I reach for the sweater on the chair near my bed and step into shoes. I grab the phone and wait for the security company to call. When the phone rings, I answer and go through the drill of identification. The system shows glass breakage on the first floor, the west window, the monitoring rep says. "Are you okay?"

"So far," I say, my ears straining to make out footsteps downstairs. The upstairs door leads to the street. The downstairs door leads to a field and the lake. I'm standing inside my bedroom door. Part of my brain is working out whether to position myself between the open stairway and the outside door or stay where I am, not far from a window to the deck. I'm leaning toward heading for the front door, where I keep my jacket—gloves in one pocket and a hat in the other—on a bench.

"The motion detector doesn't show any motion inside," the man says, "just glass breakage from downstairs. Shall we call the sheriff?"

"Yes," I say. "I can't tell whether anyone's inside, but I want someone to check." As soon as I hang up the phone, I put on warm socks and wool slacks. As I wait, I remain inside the bedroom. I've decided to stay put. In one hand is the phone, in the other is my bag, with cell phone, Filofax, and car keys.

When the sheriff's deputy arrives, he walks through the house and checks every room and all the closets. No window is broken, he informs me. My alarm system is so sensitive, even an attempt must trigger it. When I moved here, I'd briefed both the sheriff's department and the police department in town. I wanted to make sure they took any call from me seriously. This isn't the first time an officer has come into the house to check out an alarm. I suspect it won't be the last.

After satisfying himself that no one's in the house but us, the deputy goes outside to walk around the property. When he comes back in, he tells me there are footprints outside the window downstairs, but he can't tell how fresh they are. It's bitter cold, but it's a still night, and it hasn't snowed for a while. Then he adds, "I was hoping to find someone inside."

"Outside?" I ask, thinking he misspoke.

"Inside," he says. "It's been a quiet night, and I was hoping for some action."

"Do you know I'm being stalked?" I say. "It's supposed to be in your records."

"Yeah, I heard," he says. His face is thick. Like his words.

"Do you have any idea what a stupid thing that is to say?" I ask. I say it for the principle of the thing. I know he doesn't get it. I also know my words might jeopardize the response to the next incident. These are good-ole boys who spread the word. But

I can't stop myself. "You say that as if it wouldn't much matter whether I was dead or alive as long as you had fun getting an intruder."

"I didn't mean that," he says.

"Yeah, well, it *could* mean that," I say.

Now I feel as if there *is* an intruder in the house. I can't get this deputy out of my home quickly enough. "Thanks for coming," I make myself say. I may need this guy again.

As he leaves, I lock the door. Then I turn off the lights and stand in the kitchen window. I watch his car drive away, the headlights sweeping over high mounds of crusted snow. I walk around the inside of the house, window to window, door to door, checking all the locks, listening for steps, watching for shadows. I have no way of knowing whether a man, a deer, or a cougar set off my alarm.

Perhaps this has nothing to do with Paul. It's quite possible, but what I'm left with is always the same: he has bled the assumption of innocent coincidence right out of my life. His final gift to me is a life of not knowing, not ever knowing for sure.

Chapter 69

I'm in the city on business. These days I come back to the place I called home for years only when I can't avoid it. I manage my work in such a way that I'm not required to be where I don't feel safe. But when necessary—every month or so—I drive the distance. Only a few hours away, the city feels like it no longer has anything to do with me.

Just as I did in the last years I lived here, I avoid many neighborhoods, and I remain on high alert during these visits. I don't stay in the same hotel or with the same friends twice in a row. I alternate restaurants. I drive out of my way to bookstores and to coffeehouses I never frequented before.

One afternoon, between a client meeting and dinner with friends, I drive to a bookstore with a café in a suburb half an hour from my hotel. I mention my plans to no one. When I get there, I check to make sure it's as I remembered: two entrances. Yes. I scan the room and see no one I know. I get a cup of tea and wander down the aisles. About an hour after I arrive, I suddenly feel less easy than when I first walked in the door. Moving to a section of the bookstore that allows me a full view of the room, I position myself behind a pillar.

And then I see him, standing just inside the front door. He's not at all as I remembered him: solid, with broad shoulders, thick legs, and a square face. Instead, he looks fleshy and unkempt, not the meticulously groomed man I met twelve years ago. I've heard from friends who have seen him that he looked scruffy, ill even. I see what they meant.

Now it's Paul who seems to be surveying the room. He hasn't seen me. I'm sure of it. I move behind a tall shelf of books. As a feeling of frenzy rises, I will myself to remain calm and focus on leaving. Placing the books in my arms on an open shelf, I search for a route to the side door that will keep me hidden from his view. If I turn right outside the door, I think, and walk around the block behind the building, it's possible he won't see me leave. As casually as I can manage, I edge my way out.

Only when I've driven miles away, constantly looking in my rearview mirror to assure myself that I'm not being followed, does my body free my brain to think of something other than whether he's behind me. Perhaps this is simply a matter of horrible timing, a true coincidence. If I'd arrived an hour earlier or he'd arrived an hour later, he would have missed me. But it doesn't much matter. Even a chance sighting could set off a new round of stalking.

Nothing has changed. He can pay for surveillance any time he chooses. And if he's lucky, he can simply stumble into me, free of charge. Either way, I'm always left to wonder what might come at any given moment, should he decide to turn my way.

The only thing that's certain is this: I can never return to life as I once knew it, and I'll never know the life I might have had.

Epilogue

SEPTEMBER 2007

The afternoon sun hangs low, lethargic—as if it has nowhere better to be. I'm standing at the kitchen window making a pot of tea when the telephone company's truck finally pulls into the driveway. I haven't had phone service since early Monday morning. It's now Wednesday afternoon. An inconvenience, of course, but I must be the only one of hundreds of customers without service who's relieved to hear that it's due to a cable cut during nearby construction.

I turn off the alarm system and meet the technician at the door. He just wants to make sure my phone is working now, he tells me. It was still dead a half hour ago, so I ask him to wait at the door while I find the portable phone. Within seconds I'm back. Yes, indeed, I assure him. I have a dial tone.

"A cut cable?" I ask, mostly as a formality.

"Well, yes," he says, and then proceeds to tell me that my line had an additional problem.

I feel dizzy, but just for a second. "Another problem?"

In the box up at the highway, my line was cut, he says, and reattached to another line.

"Okay," I say slowly. "And how do you think that happened?"

Most likely a rodent, he says. It happens sometimes. The cut, he clarifies, not the reattached line. He can't figure that one out.

In answer to my subsequent questions, I discover that the box holds about two hundred lines bundled close together and that mine was the only one cut, which is disturbing, of course, but not nearly so much as the fact that it was then reattached to someone else's line.

"Could it have been done deliberately?" I ask.

It's possible, he replies, but who would want to do that?

As if I've brushed against a live wire, a jolt of anger rushes through me, but I tamp it down. I explain that I've been stalked and things like this have happened in the past.

Does your stalker work for the phone company, he asks, because otherwise it would take quite a while to figure out which line is yours. Besides, he repeats, what kind of person would do that?

I don't take time to detail the kind of sociopath who would do such a thing, nor do I suggest how easy it would be to offer someone money to take the time. Instead, I thank him, lock the door, and reset the alarm system.

Two nights later, at dinner with friends, I tell them the story. Ben, whom I've known since the first grade, says, "*Right*, a rodent. What a clever squirrel it would take to not only chew through your line and leave the rest alone, but then to attach your line to another. *Right*, a rodent."

"Exactly!" I say, nearly melting with relief. Clever squirrel indeed.

I spend the rest of the weekend at home, alone. I sleep fitfully. I keep the windows closed. Leaving even a crack open to the outside world—seeing anyone, talking to anyone—feels like more than I can take. I retreat into the calm of my books, where nothing outside their covers can touch me.

By Monday morning, I feel stronger, but still off balance. To help right myself, I order a new cell phone. Changing my cell service may not be logical. It may not make me safer. It's simply the only thing I can think of that will help me get past the whole incident.

On Wednesday, I get a call from a customer representative at the wireless provider wanting to verify that my new phone is supposed to be delivered to an address in an adjoining state. I thank him for checking and give him the correct address. He informs me that they already have that one, but that someone added a new shipping address to my file. When I ask how this could have happened, he has no explanation. "But we thought we'd better check since it's a different state than your billing address."

I don't know anyone in the town he names, but I think I recognize it as being near a cabin Paul had several years ago.

What are the odds?

As determined as I am not to let this get to me, it wears me out. Mostly, I suspect, because of all the energy I expend not giving in to my rage. Perhaps if the stalking had ended, I'd give in to it. I don't know. All I know is that I'm afraid to let the slow-burning fire inside me run out of control. I still need a cool brain because my whole life—every minute of every day—is dictated by one current: whether or not I'm safe.

The only time I'm tempted to give in to my anger is when I think of all the other things I could have done with the energy these efforts have cost me. Every time I packed up a house, I could have been writing. Every time I settled into another neighborhood, I could have been involving myself in a worthwhile local group, or meeting a future lover. That's what I've sacrificed. And there's no way of getting that time back.

Not that I think of these as completely lost years. I've done my best to turn them into something worthwhile. I've learned to find comfort in the small things: the scent of lavender on clean sheets, the sound of a loving voice on the phone, a day without incident, a heavy fall of snow, or the luscious sound of rain. A pile of books at my bedside assures me that isolation need not mean loneliness. Photos of family and friends remind me that they need not be with me to be present.

I see my mother and my grandmother in my fingers and in the veins on the back of my hands, and I know I have their capacity for joy, and survival. I cup my face in my hands and feel my father's cheekbones and jaw, and I know I have his clarity of mind, and courage.

Mostly, I've learned to shed the damage while protecting my essential self, which leaves me wiser than before and open to possibility.

But I can't help wishing I could have grown through another means, because to be the eternal target of a maniac, to have to live the way I do, to know that even if he were caught, he'd probably be more dangerous still—that is madness itself.

So, to pretend that I'm fine and life is good would be to lie.

Here are the facts: as long as Paul lives, I believe he will stalk me—not continuously, but often enough to remind me he can. Because of him, I've given up any semblance of a normal life. I've given up homes, neighborhoods, and whole cities. I've given up work, friends, and possibilities of love. I've even given up my name.

Some days, I almost give up hope.

But here's another fact: Paul made a strategic error when he chose me as his target. By picking me, he denied himself his greatest desire: to conceal his true self. The longer the stalking continues, the longer the list of those who know what kind of man he is. The more he interferes with my life, the more completely he blows his cover. And what the detectives said ten years ago still holds true: if anything happens to me—or to anyone in my family—anywhere in the world, he'll be their only suspect.

More than anything, stalkers like Paul need to obliterate a person's spirit, and I want to believe that no matter what he does, he will not have done that to me. But even if he does, what will he have won? What exactly do you get when you destroy the life of someone once foolish enough to love you?

What you don't win is someone's silence. In the end, whatever else happens, I have told my story: the story of an ordinary woman who was victimized, but was not a victim—a woman who suffered, but was not defeated.

He may have moved me to the margin of life, but he can't erase my words.

ACKNOWLEDGMENTS

Like many Irish, I was raised to be an optimist, no doubt a defense against our natural inclination toward melancholy. A higher standard than the simple stoic edict to buck up and move on, the Celtic call requires one to actually be *grateful* in adversity. Things could always be worse. In fact, things *are* worse. For *someone. Somewhere.*

We're meant to feel lucky, even when we're not. And when we *are* lucky, we're meant to appreciate every bit, so there's no limit to our gratitude. I regularly tell myself how fortunate I am to have survived a man like Paul. And I *have* survived. I have a scar to prove it. On the soft flesh of my inner left arm, halfway between my wrist and my elbow, I'm branded—with a double-dash scar, like the M in Morse code. Dash–Space–Dash. The dashes are white and slightly raised. The space between them is my natural color, pink, unblemished. All I have to do to remember is to run the index finger of my right hand over the two white marks that form the faint, raised scar on my veined flesh. M as in the brand left by the HIV test I took right after I left him. M as in near Miss. M as in it's dangerous not to be Mindful.

And the fact is, I've never been more mindful, of the friends and family who love me, for a start, without whom I could never have dared to write this book. I thank them for their gifts: affection, honesty, loyalty, protection, humor, and countless hours of ordinary life. That is the stuff of love, and sanity. For every minute of it, thank you. And to those who read versions of this manuscript, encouraged me, and offered good counsel, an extra thank-you.

My agent, Marly Rusoff, saw promise in an early manuscript and understood that the last thing I wanted to do was sensationalize my story. With sound and wise counsel, she encouraged me, made valuable suggestions, and led me to the right editor and publisher. She and Michael Radulescu never forgot my need for safety; I will always be grateful for their efforts, laced with kindness and good humor.

Only the most fortunate writers have the chance to work with a gifted editor, the kind who conducts the writing-editing process like a master class. My editor, Jennifer Barth, is one such gifted editor. This was a painful story to tell, and, were it not for Jennifer's patience and always spot-on guidance, my story may have been published, but in an inferior form. She held my spirit high while helping me find the best shape and the perfect pitch. And then, of course, she made sure there wasn't a lazy word or a false note in sight.

Thank you also to Christine Van Bree, Mark Jackson, Tina Andreadis, Leslie Cohen, Jeanette Zwart, Christine Boyd, Doreen Davidson, Brad Wetherell, and everyone at HarperCollins for the generous enthusiasm with which they launched this book.

Finally, although they are no longer here, I want to thank my parents. My mother taught me to love—and work at—language. As my first fan and first critic, she taught me to value a reader who compliments a fresh image and underlines a weak passage with equal vigor. Until her last breath, she applied those same rules to character, too. My father showed me how to tell a story straight on. He recognized that a story that doesn't own the pain as well as the triumph is only half a story.

Not long before he died, he said, "It's just a matter of time before you write about us. Would you consider changing the names if I'm still alive?"

Well, Dad, here you are.

Insights,
Interviews
& More . . .

Living Under a Pseudonym

I'M IN THE LIBRARY doing research when I realize that it's nearly two o'clock, time to call the radio producer for my pre-interview. I slip roll 38 into place in the microfilm drawer and head for the parking garage. A few minutes later, I find a spot on a quiet street a few blocks away, under the shade of an oak tree.

I've taken to wearing a particular necklace when I'm meant to be Kate Brennan. I finger the smooth oxidized silver of the links and the cool rounded glass of the locket, as if this simple action will transport me. Actually it grounds me, which may sound foolish, but writing under a pseudonym, and living under another name, doesn't become second nature when your first nature is to be authentic. It's difficult to keep from automatically and absentmindedly offering your real name when you introduce yourself. And you'd be amazed at how easy it is to make a slip about where you live in a casual exchange about the weather.

"Kate Brennan, Kate Brennan, Kate Brennan," I repeat as I tap in the number my publicist gave me earlier in the week. Ready, set, go: "Hi, this is Kate Brennan," I say to the producer. This is the drill. When someone requests an interview, my publicist arranges for me to make the call. That way I can either use a

"Writing under a pseudonym, and living under another name, doesn't become second nature when your first nature is to be authentic."

public phone, or my cell phone, which has the number blocked as a "private caller."

Driving back to the library, I feel myself relax, as if my shoulder and back muscles know that for the rest of the day they won't have to steel themselves to guard my pseudonymous life. It's an unpleasant reminder of the past. These days, I live under my own name and write under another, but back when the stalking was at its worst, I was forced to adopt different identities to protect myself. It's what the police advised. At one point, I signed a lease for a town house under one name while using another for the cable and electrical companies. I traveled under yet another name, backed up by a credit card, and a letter from local sex crimes detectives detailing the situation.

You'd think all this identity-swapping would have prepared me for writing under a pen name, but it didn't. Proud of how open I've been in this memoir, I feel almost dishonest hiding behind yet another name. Perhaps it's because I don't like to lie. And I'm not good at it. When I had to create a backstory for the "me" who lived in the town house, I never got used to deceiving my neighbors. Even though it was necessary, leading a double life never felt good. Not to mention how exhausting it was to remember—or make up—stories every time I met someone new.

So, then, why write under a pseudonym? Writers have many reasons for not using their own names. For centuries, it was the only way women could have their work taken seriously. Think of Acton, Currer, and Ellis Bell (Anne, Emily, and Charlotte Brontë); George Eliot (Mary Anne Evans); George Sand (Amandine Aurore Lucile Dupin). Even now, sometimes initials mask an identity—J(oanne) K(athleen) Rowling and S(usan) E(loise) Hinton, for instance. I think Charlotte Brontë captured it best: "I am neither a man nor a woman but an author."

Today, some writers use a pen name as a means of switching genre rather than gender (Benjamin Black/John Banville)—or, put another way, of unleashing different parts of themselves (Rosamond Smith/Joyce Carol Oates and Barbara Vine/Ruth Rendell). Some writers use a pen name for privacy. It's a way to separate your everyday life from your literary life. (I'd feel like a spoiler if I gave contemporary examples.) ▸

Living Under a Pseudonym ***(continued)***

While the result of my pseudonym may be privacy (a relative state since I feel as if I've spilled my guts on the page), its roots are more fundamental. The goal of my subterfuge is safety. My stalker is still at large; writing under a pseudonym seemed the only way to tell my story as fully as I wanted without exposing myself to possible repercussions. I wanted my story to be as true to the terror of being stalked as possible. I worked hard to capture every detail, every memory, as precisely as I could. And then, I not only changed *my* name, I changed his name and every name in the book (even the dog's).

Other than that, the only details I altered were the ones that would clearly identify my stalker. Even so, I expect "Paul" to eventually recognize himself on the page. I'm hoping he registers that identifying himself would simply implicate him further. Yet for all I know, it will enrage him and put me at greater risk. In other words, I have to be prepared for anything.

This, of course, raises the question: Why write a memoir at all? Aside from being tired of keeping my life—and my work—so far under the radar that I dropped off the screen, I wrote because I *could.* I realized that I had the ability to give voice to not only my story but to those of millions of other women who were afraid—or embarrassed—to reveal the reality of living at the whim of a maniac. Fictionalizing my story under my own name would have diminished, even denied, that reality. (Besides, as a writer friend reminded me: "You couldn't make this stuff up.")

Keeping silent about my experiences would have helped keep the myth alive that being stalked (or raped or abused in any way) is something shameful, that it's somehow the victim's fault. My hope is that telling my story will free millions of other voices and that the cumulative affect will help shatter that myth.

While the act of writing this book has in many ways freed me, the irony is that its publication makes me both safer, broadening the field of people who know and can help protect me, and more vulnerable, as I must rely on so many more people to keep my identity hidden. When nearly every cell phone has a camera, potential exposure lies right outside your front door and inside every room you enter. And then, of course, too many people today are eager to reveal everyone else's secrets and post them for the entire world to see. It's frightening to think someone might make a game of trying to find out who I am. (I try not to dwell on this.) Because it's so much more than a matter of convenience or privacy, I depend on all my readers—friends and strangers alike—to be good-hearted and decent. (I'm not naïve, just optimistic.)

Still, after years of writing under my own name, I feel a bit cheated to have to hide yet again behind another one. Faced with selecting my pseudonym, I thought back to a day, at the age of twelve, when I was eager (or so I thought) to trade in my name. I admonished my mother for giving me such a plain name. When she asked me what I would prefer, "something exotic" was all I could manage. "When you're of age, be my guest, but in the meantime, we'll call you by your own, which is a beautiful name," and she sent the simple but lyrical syllables of my first and middle names into the kitchen air.

When it came time to actually choose a pen name, I was driven by a need to have it feel *connected* to me. I amazed myself with how complicated I made the process. I rearranged the letters of my first, middle, and last names. The results all sounded like strippers. Next I combined my initials with those of my parents. Those names all sounded like they belonged to eighteenth-century poets. Then I tried variations with my confirmation name, but since I didn't like that one even when I chose it, the results held no appeal.

Finally, running out of time, I opted for a name that was not obviously linked to me, yet one that *felt* familiar. I settled on Kate Brennan, a name as clean and simple as my own.

Now that I'm attached to it, I can only ask that no one rob me of the protection it affords.

The foregoing essay appeared online at Powells.com, the Web site of Powell's Books.

Writing *In His Sights*

"Was it cathartic?" That's the most frequent question I hear since writing *In His Sights.* The eyes behind the query generally share a common reflection: they're looking for a yes.

I always disappoint. Though I take a few seconds each time to see if my answer has changed, it remains no.

In one way, being stalked is like being abused or being raped. In the aftermath, you are never the same. You never get over it. You may figure out a way to live through it, to survive it. You may even figure out a way to get on with your life, but you never drop your vigilance. You are never purged of the memories, and never again will you feel entirely safe.

Somehow I knew that, so I didn't expect writing my story to be cathartic. Why, then, write about the experience—and relive each terror and trauma in every step of the rewriting and editing—if it will not offer relief?

In the moments, often stretching into days and weeks, when it would have been easier to stop writing, to stop remembering, a single thought kept me at my desk: Millions of people (mostly women) have lived, are living, or will live through a version of my story. And most of them, not professional writers, do not have the ability to write their stories for publication.

I have had many advantages in life. I was born into a well-educated, articulate

"In the moments . . . when it would have been easier to stop writing, to stop remembering, a single thought kept me at my desk: Millions of people (mostly women) have lived, are living, or will live through a version of my story."

family, and I had parents who provided each of their children with an excellent education and helped and encouraged us to develop our natural gifts—in my case, writing and researching. They imparted these benefits with a single, clear dictum: Use the gifts you've been given, and not just for your own gain. Your work must matter in the world outside your door.

When the pain of writing felt as if it would nearly destroy me, when it felt more like drinking toxins than cleansing myself of them, I felt an obligation to continue not just for myself, but for every woman who cannot give voice to her own story.

Something else keeps women who are stalked silent. Though our stories differ in the details, most of us walk around with the burden, heavier than our fear and anger, of the shame we feel for being in such a situation in the first place. It's hard to shake the feeling that we somehow brought this on ourselves, especially if we're being stalked by someone we once loved.

This, of course, prompts another obvious question: Why don't women leave the minute they see the clues to a man's abusive and controlling character? More often than not, the answer lies in personal history. A family steeped in addiction made me vulnerable to a man such as this; my (hard-won) resilient nature led me to believe I could help—even change—him. And because I met him at a time when he was faced with an unfathomable family tragedy, the idea of giving up on him seemed beyond heartless.

Not all women who are stalked are stalked by men they know, though most of us are. Not all of us are left with an open-ended scenario, though most of us are—only an insignificant minority of stalkers is ever prosecuted or convicted. Some women, like me, choose to live alone. Others go on to have families, and children, which makes the threat and danger far worse.

Whatever the specifics of our lives may be, our feelings of shame, logical or not, are exacerbated each time friends, family members, or legal authorities throw us a look or a comment that barely disguises (if at all) the idea that we, the victims, are the crazy ones, that we're either exaggerating or imagining things. And if not that, they suggest that we should be flattered by what they mistake for love or attraction. Making matters even worse is every movie that either trivializes or sensationalizes the crime ▸

Writing *In His Sights* *(continued)*

of stalking rather than revealing the trauma of our every waking hour. It's no wonder we tend to keep our stories to ourselves.

In the end, what drove me to write *In His Sights* and kept me going was my desire to speak, not only for myself, but for every stalking victim—daughter or sister, mother or friend, stranger or neighbor—who stands before me, beside me, and ahead of me, and who hasn't the means or the confidence to publicly tell her story.

Curiously and surprisingly, deciding to break my own silence by writing this book has turned out to be freeing, which isn't at all the same as cathartic. I'm no longer held captive by the feeling that I'm harboring a dirty little secret, the secret of being the target of a madman.

It is my hope that telling my story offers freedom to every woman with a similar story of her own.

A Conversation with Kate Brennan

Many people can't begin to fathom what you have gone through and continue to endure, yet you're still here and still fighting back. What message do you want to convey with this book?

I want my story to inspire women to see that no matter how bad a situation may seem, there's almost always hope. Beyond family and friends, many people—police, therapists, support groups, organizations—can help. Staying in a relationship that is unhealthy and damaging is tantamount to choosing a slow and painful death. Even with the stalking, leaving Paul was far better than staying with him.

When celebrities are targeted by stalkers it becomes headline news, but most Americans are uninformed about how many ordinary women in this country are victimized, too. What do statistics reveal and what can be done to raise awareness of this often deadly phenomenon?

One in twelve women will be stalked in her lifetime; each year, more than a million women are stalked in the United States alone. The majority of the time the stalker is someone she knows. Just over ▸

half of stalking victims report the crime to police; only 13 percent of stalkers who are charged are prosecuted, and of those, just over half are convicted. Tally up all these numbers and you get a pretty discouraging picture.

It's my hope that my book will not only heighten awareness of stalking, but also lead to greater support for the groups and organizations that help victims of abuse and violence of all kinds. Since 1990, every state has passed an antistalking law, and now there's a federal law as well, but there is still much more that could be done.

How have detectives in the sex crimes unit been able to help you, and how have their hands been tied?

While we were ultimately unable to bring build a case for prosecution, the detectives helped enormously in several ways. For starters, they validated my experience. While this may sound like a small thing, you'd be surprised at how many otherwise intelligent and sensitive people dismiss this crime as "love" or suspect the victim of exaggeration. Perhaps even more important, they offered very specific advice about how to increase my safety. And finally, they let my stalker know, in no uncertain terms, that if anything happened to me—anywhere in the world—he'd be their only suspect. That warning spoke volumes: even if they couldn't prove the stalking in a court of law, they knew he was the perpetrator.

Is the legal system doing everything it can to protect women who are being stalked?

Even though we now have antistalking laws, not all the enforcers of those laws—cops, prosecutors, judges, juries—are knowledgeable about the crime itself. Some discount the danger or blame the woman. As long as there are people in these positions who have that attitude—whether it's stalking, rape, or domestic or sexual abuse—women won't receive the protection they deserve.

At a certain point you had to move twenty times in eighteen months. Can you explain what it's like to live on the run, constantly looking over your shoulder, suspecting that every stranger you see might have been sent to harm you, and always feeling vulnerable?

Being stalked has left me more suspicious than I would like to be and more cautious than is natural in ordinary circumstances. Living under—and writing under—a false name means living a double life. In each new situation—on a plane, at a meeting, in a hotel, at a party, in a coffeehouse—I have to stop and think: Can I trust this person? What name do I use? What is safe to reveal? It's exhausting. You have to have a strong sense of yourself to stay balanced and to forgive yourself when you are too cautious, or worse yet, when you are not cautious enough.

Where have you found the strength to keep going during this terrifying ordeal?

I'm by nature a resilient, optimistic person, and I was taught that in times of difficulty you figure out how to cope, how to solve problems, and how to count your blessings.

What was your rock-bottom moment?

When my stalker moved into a house across the street, I gave up my apartment, put everything in storage, and left the country. For the first time in my life, I despaired, not that the stalking would never end, but that I wouldn't survive with my spirit intact. Yet, even in those worst days, I believed in my own resiliency, and in the kindness of good-hearted people, some close, some strangers.

Do you think you've reached an understanding of why you were drawn to Paul in the first place? ▸

All I have to do is look at my mother and her mother, both smart, strong women who married smart, strong men who were addicted to alcohol (in my father's case) and prescription drugs (in my grandfather's). I thought a loving woman stayed with a man no matter how hurtful or abusive he became. My mother and grandmother loved men who at the core had a kind heart and a strong sense of right and wrong. (My father ultimately surmounted his addiction and did his best to make up for the pain he had caused others.)

It took me a while to see that the addict I chose had neither a good heart nor a conscience.

From what you've learned from experts and in your own research, are there certain characteristics (e.g., narcissism, entitlement, rage) found in most stalkers?

The vast majority (more than 80 percent) of stalkers are white males of above-average intelligence and income. In general, stalkers tend to be narcissistic, controlling, highly manipulative, and deceptive. They view themselves as victims, so they don't take responsibility for their actions and feelings, and they blame their victims. They are unusually jealous and unable to cope with rejection. Many stalkers have personality disorders.

Is the duration and intensity of Paul's campaign against you unusual or common?

The average stalking experience lasts approximately two years; my situation started up fifteen years ago. Paul has enough money and free time to keep his harassment going as long as he wants. And I'm particularly threatening to him because I know so much about him: I was a close friend of his aunt's and met him at a time when his family's secrets were being publicly revealed.

How have your friends and family reacted to your plight?

Some family members and friends believed me from the start and did whatever was in their power to help me. Others held me at a distance, as if being allied with me too closely might reflect on them negatively, and still others turned away from me completely. Some people just can't face things they fear or don't understand. One of the most wonderful things to come out of writing this book is that I've reconnected with one of my brothers. We used to be close, but drifted apart. After reading my book, he told me how sorry he was that he wasn't there for me during this horrible time, and how much he wanted to be part of my life in the future. He's such a loving brother and friend. I will always be tremendously grateful for this.

Have you had to cut individuals out of your life for not believing you, or worse, exposing you to danger?

Yes. At first it was extremely painful, but in time, I found it easier to narrow my circle of friends because it felt essential to my safety and sanity. As with any significant life event, we learn who our true friends are in times of crisis.

Looking back on your relationship with Paul, are there certain incidents, behaviors, or revelations that might have been red flags?

His refusal to talk openly about his past was a huge red flag. Our relationship was complicated by the fact that right after we met, there was a tragedy in his family. I interpreted his secrecy, silence, and withdrawal as symptomatic of the difficult time he was going through, and as temporary. Even as his behavior grew increasingly suspect, I told myself that I couldn't abandon someone who was in such pain. Growing up in an alcoholic home gives you a high tolerance for harmful relationships.

Do you have hope for a more normal future? ▸

I've been warned that as long as he's alive, Paul is apt to mess with me just often enough to remind me he can, so living the way I do will probably always be, to some extent, my "normal." Since I'm a realist as well as an optimist, the most I allow myself to hope for is that I outlive him by many, many years.

If you could say anything to Paul right now, what would it be?

This is going to sound trite, but the only thing I'd say is, "Get help." I have no need to make him feel remorse for what he's done to me. I feel that disconnected from him.